Role of ICT for Multi-Disciplinary Applications in 2030

RIVER PUBLISHERS SERIES IN COMMUNICATIONS
Volume 47

Consulting Series Editors

ABBAS JAMALIPOUR
The University of Sydney
Australia

HOMAYOUN NIKOOKAR
Delft University of Technology
The Netherlands

MARINA RUGGIERI
University of Rome Tor Vergata
Italy

The "River Publishers Series in Communications" is a series of comprehensive academic and professional books which focus on communication and network systems. The series focuses on topics ranging from the theory and use of systems involving all terminals, computers, and information processors; wired and wireless networks; and network layouts, protocols, architectures, and implementations. Furthermore, developments toward new market demands in systems, products, and technologies such as personal communications services, multimedia systems, enterprise networks, and optical communications systems are also covered.

Books published in the series include research monographs, edited volumes, handbooks and textbooks. The books provide professionals, researchers, educators, and advanced students in the field with an invaluable insight into the latest research and developments.

Topics covered in the series include, but are by no means restricted to the following:

- Wireless Communications
- Networks
- Security
- Antennas & Propagation
- Microwaves
- Software Defined Radio

For a list of other books in this series, visit www.riverpublishers.com
http://riverpublishers.com/riverpublisher/series.php?msg=Communications

Role of ICT for Multi-Disciplinary Applications in 2030

Editors

Leo P. Ligthart

CONASENSE, The Netherlands

Ramjee Prasad

GISFI, India
CTIF, Aalborg University, Denmark

Published, sold and distributed by:
River Publishers
Niels Jernes Vej 10
9220 Aalborg Ø
Denmark

River Publishers
Lange Geer 44
2611 PW Delft
The Netherlands

Tel.: +45369953197
www.riverpublishers.com

ISBN: 978-87-93379-48-0 (Hardback)
978-87-93379-47-3 (Ebook)

©2016 River Publishers

Contents

Foreword

The theme of this book is "Role of ICT for multi-disciplinary applications in 2030", which is absolutely appropriate to explore with regard to the CONASENSE vision of looking at services utilizing the Communications, Navigation, Sensing, and Services (CONASENSE) paradigm in a period of 20–50 years from now. The vision of CONASENSE society to bring about active integration of the three worlds of communications, navigation, and local/remote sensing – that have been apart for years requires a multidisciplinary approach.

This 4th Communication, Navigation, Sensing, and Services (CONASENSE) book brings together contributions from another society, namely, Global ICT Standardization Forum for India (GISFI). The Global ICT Standardization Forum for India (GISFI) is an Indian standardization body active in the area of Information and Communication Technologies (ICT) and related application areas, such as energy, telemedicine, wireless robotics, and biotechnology. GISFI is an effort to create a new coherence and strengthen the role of India in the world standardization process by mapping the achievements in ICT in India to the global standardization trends. Further, GISFI is focused on strengthening the ties among leading and emerging scholars and institutions in India and the world; to develop and cultivate a research and development agenda for the field.

The book is composed of nine chapters; with four from CONSENSE and five from GISFI-allied contributors.

Chapter 1, "Multi-disciplinary Applications of Wireless Sensor Networks: Challenges and Research Directions", discusses cutting edge technologies in relation to wireless sensor networks and includes: cognitive radio, communication relying on orthogonal frequency division multiplexing and physical layer security, and context aware applications relying on wavelet technology.

Chapter 2, "Multi-disciplinary Applications Requiring Advanced IoT and M2M", discusses the fundamental differences between M2M and IoT based on drivers and enablers that drive the commercial utility of these technological aspects. The chapter also elaborates some of the multi-disciplinary applications based on M2M and IoT.

Chapter 3, "Experimental Activities to support Future Space-based High Throughput Communication Infrastructures" describes the need for high-throughput satellite (HTS) communications operating in the mm-wave bands as part of a global broadband ICT in future. HTS allows for the globalization of gigabit/terabit data communications capabilities, improved reliability of integrated ground-based and space-borne networks, and improved performance compared to present generation solutions".

Chapter 4, "High-capacity Ground, Air, and Space-based ICT Networks for Communications, Navigation and Sensing Services in 2030", proposes an integrated terrestrial plus satellite system to provide the capabilities of communication, localization, and sensing for the innovative multidisciplinary services of the future pervasive Internet "Ecosystem". Services via such networks enhance the quality of our everyday life as well as applications in critical situations and environments. The network architecture affords remarkable quality and performance needed to satisfy the requirements of the future society based on the Internet of Things.

Chapter 5, "Organizing International ICT Research for Multi-disciplinary Applications", is written from the vision that weaknesses in present ICT can be overcome by future multi-disciplinary ICT research. Based on foreseeable trends in ICT the chapter presents a research vision which necessitates activities by all stakeholders. Furthermore, research and education in future ICT should be international, multi-disciplinary and on applications while standardization is important to be involved.

The focus of Chapter 6 is on "Convergence of Secure Vehicular *ad-hoc* Network (VANET) and Cloud in Internet of Things". VANET aims to increase quality, safety, and security, as well as to reduce cost and complexity. Realization of such network needs integrated efforts from automotive, software, networking, and telecommunications industries.

Chapter 7 describes an innovative network called "Heterodox Networks: An Innovative and Alternate Approach to Future Wireless Communications". The accumulation of subscribers and their movements creates an apparent capacity demand which is introduced in the chapter under the name 'Place Time Capacity (PTC)'. To combat PTC the Heterodox network combines two complementary systems, one is based on a fixed infrastructure and the second has a dynamic infrastructure.

Central theme in Chapter 8 is: "Network neutrality for CONASENSE Innovation Era". Network Neutrality (NN) aims that every end user has the equal right to access the internet and use the legal internet content and applications. In this chapter, it is motivated and illustrated with some futuristic scenarios why it is important to understand the impact of NN on CONASENSE services.

Chapter 9, "CONASENSE at Nanoscale: Possibilities and Challenges", elaborates the e-challenges in communications, navigation, sensing, and services at nanoscale relying on nanotechnology and gives a vision on the wide range of possible applications at nanoscale. Use of CONASENSE at nanoscale is manifold; in the chapter some examples in the home, or shopping environment and in medical treatment get attention.

We are certain that the book provides a rich and interesting coverage of diverse aspects concerning services merging the three worlds.

Leo P. Ligthart
Chairman CONASENSE,
The Netherlands

Ramjee Prasad
Founder Chairman, GISFI, India
Professor, Director, CTIF, Denmark

List of Figures

List of Tables

1

Multi-Disciplinary Applications of Wireless Sensor Networks: Challenges and Research Directions

Homayoun Nikookar[1] and Leo P. Ligthart[2]

[1]Netherlands Defence Academy
[2]CONASENSE Foundation, Delft, The Netherlands
E-mail: h.nikookar.nl@ieee.org; leoligthart@kpnmail.nl

Abstract

ICT is the key technology for Wireless sensor networks (WSNs). Wireless sensor network is a network of low-size and low-complexity devices that sense the environment and communicate the gathered data through wireless channels. The sensors sense the environment, and send data to control unit for processing and decisions. The data are forwarded via multiple hops or connected to other network through a gateway. *WSNs are applied in a wide and multi-disciplinary range of areas from monitoring environment and surveillance, to precision agriculture, and from biomedical to structural and infrastructure health monitoring.* Technological advances in the past decades have resulted in small, inexpensive, and powerful sensors with embedded processing and radio networking capability. Distributed smart sensor devices networked through the radio link and deployed in large numbers provide enormous opportunities. As will be explained in Section 1.2, these sensors can be deployed in the air, on the ground, in the vehicles, in or around buildings, bridges, or even on the patient's body.

ICT in general and Wireless communications in particular are the prominent technology for WSNs. In this chapter, important enabling technologies for the development of future WSN will be discussed. Cutting-edge technologies such as Distributed Beamforming (DB), Cognitive Radio (CR), Joint sensing and communication with Orthogonal Frequency Division Multiplexing (OFDM), WSNs Physical Layer security as well as Wavelet technology for

Role of ICT for Multi-Disciplinary Applications in 2030, 1–22.

context-aware WSNs will be described and their importance in the future developments of WSNs will be discussed.

1.1 Introduction

A typical sensor node in a wireless sensor networks (WSN) has a small size (at the size of a button) which has sensors, radio transceivers, a small processor, and a memory and a power unit. The small size is the focus of this chapter but is a limitation as well. With the proliferation of WSN, the requirements on prime resources like battery power and radio spectrum are put under severe pressure. In a wireless environment, the system requirements, network capabilities, and device capabilities have enormous variations giving rise to significant design challenges. This is obvious from examples such as wireless ad hoc networks, mobile cellular heterogeneous networks, WiFi, and ZigBee or wireless short-range personal area networks. More specifically, in vehicular networks, for instance, the imposed safety requirements, low packet delivery rate arising from congestion in dense traffic networks, and high bit-error-rate (BER) due to Doppler shift degradation caused by high node mobility [1] are remarkable problems which challenge the design of network. Therefore, there is an emergent need for developing energy efficient, green technologies that optimize premium radio resources, such as power and spectrum, even while guaranteeing quality of service. Moreover, many WSN operate under dynamic conditions with frequent changes in the propagation environment and diversified requirements. All these trends point to flexible, reconfigurable structures that can adapt to the circumstances and the radio neighborhood. The nodes of future WSNs will most likely be context-aware and heterogeneous with energy harvesting capabilities. Wireless sensor nodes (motes) with energy scavenging capabilities in [2] and the context aware wireless sensor network for assisted living in [3] are just a few examples to name. In the following sections, major enabling technologies for the realization of this perspective on future radio sensor networks are explained.

1.2 Multi-Disciplinary Applications of WSNs

The potential applications of distributed WSN are multi-disciplinary and very broad ranging from civilian to military scenarios. In all applications, a process should be sensed or monitored remotely by distributed sensor nodes. Major civilian application areas include (but not limited to) the following fields:

- Distributed wirelessly connected sensors for earthquake sensing
- WSN for weather monitoring, hazardous weather detection, and estimation of 3D wind direction
- Air pollution monitoring with distributed and networked wireless sensors
- Climate change monitoring with Distributed wireless sensor network
- Forests and farm monitoring with WSNs
- Fire monitoring with WSNs
- Surface exploration, turbulence, and wind field measurements with distributed radio sensor network
- Surveillance and border control with Distributed radar sensor network
- Road traffic monitoring with wireless sensor network
- WSN for Smart Energy Network (Smart Grid)
- Structural health monitoring of infrastructure (dikes, dams, and bridges) with WSNs
- Distributed radio sensor network for tracking of large number of aircrafts in Air Traffic Control applications
- Application of WSNs in the agricultural and food chain monitoring and control
- Wireless sensor network for monitoring complex machinery, smart homes, manufacturing processes, and robotics
- Medical application of WSNs.

The military applications of wireless senor networks include, but are not limited to:

- Distributed network of areal sensors (including radar, camera, and other sensors) mounted on Unmanned Aerial Vehicles (UAVs) for different purposes
- Distributed passive radar sensor network for detection and tracking of low-signature targets
- Automated target recognition for smart weapons
- Guidance for automatic vehicles
- Battle field surveillance
- Surveillance of nuclear power plants
- Detection of chemical attacks
- Automated threat recognition and classification

As an example of WSNs application areas, Figure 1.1 shows the architecture of a WSN system for the precision agriculture which involves monitoring soil, crop, and climate conditions in a field, generalizing the result and providing a decision support system for treatments or taking differential action such

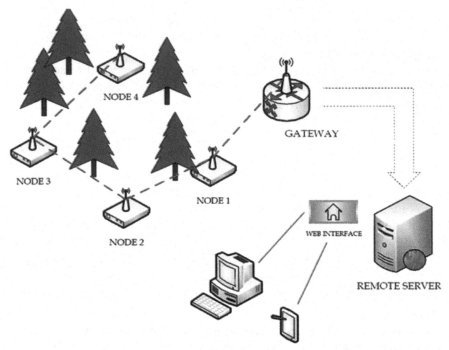

Figure 1.1   WSN scheme for agricultural application [4].

as real-time variation of fertilizer or pesticide application [4]. The network comprises a set of sensor nodes and a base station that communicate with each other and gather local information to make global decisions about the physical environment in the agricultural domain. More specifically, in [5], application of real-time smart sensor array for measuring soil moisture and temperature is reported with the sensor readings used for scheduling irrigation in several agricultural fields.

Figure 1.2 shows the application of WSNs in maritime communication. The sensors on board sense the environment [6]. The communication is via radio transmission which can be among others the intelligent WiMAX, i.e. an advanced version of WiMax which smartly adjusts its operating parameters, including power management, subcarrier selection, and spatial beam directing. Using radio communication and by combining information acquired via sensing, the final navigation decision can be made which can be useful for the situations when the Global Positioning System (GPS) is not operational or not available. Another interesting application is the extension of the coverage of WSN in the maritime domain by acquiring sensor data on board of boats

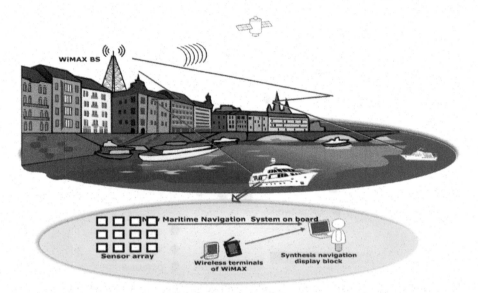

Figure 1.2 Application of WSNs in maritime communication [6].

far away from the base station by cooperative involvement of the nodes in between.

1.3 Distributed Beamforming for WSNs

In this section, we consider the application of WSNs in the maritime communication and in particular discuss how the distributed beamforming (DB) technique as a cutting-edge technology could increase the coverage of WSNs. The DB technique is extensively explained in [7].

A wireless sensor network is formed by radio nodes that are geographically distributed in a certain area, which is shown in Figure 1.3. The nodes are wireless terminals, or sensors in the network. Here, we propose to employ DB in the radio sensor network in order to form beams toward the Distant Node. The idea is equally applicable to future Cognitive Radio (CR) network where the nodes of the CR network are able to forward the signals to distant cognitive radio (DCR) node cooperatively. In the context of cognitive wireless sensor network and by adopting the DB method, the CR wireless sensor network increases its coverage range without causing harmful interferences to the already Primary radio services [or Primary Users (PU)] operating in the environment and at the same frequency band.

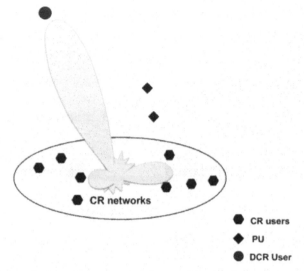

Figure 1.3   Distributed beamforming for cognitive wireless sensor network [6].

DB is also referred to as collaborative beamforming, and is originally employed as an energy-efficient scheme to solve long-distance transmission in WSN, in order to reduce the amount of required energy and consequently to extend the utilization time of the sensors [8]. The basic idea of DB is that a set of nodes in a wireless sensor network act as a virtual antenna array and then form a beam toward a certain direction to collaboratively transmit a signal. It has been shown that by employing K collaborative nodes, the collaborative beamforming can result in up to K-fold gain in the received power at a distant node. Recently, a cross-layer approach for DB in wireless ad-hoc networks has been discussed in [9] applying two communication steps. In the first phase, nodes transmit locally in a random access time-slotted fashion. In the second phase, a set of collaborating nodes, acting as a distributed antenna system, forward the received signal to one or more faraway destinations. The improved beam pattern and connectivity properties have been shown in [10], and a reasonable beamforming performance affected by nodes synchronization errors has been discussed in [11]. DB has also been introduced in relay communication systems. Different types of relays have been considered, e.g. Amplify-and-Forward (AF) relays, Decode and Forward (DF) relays. The models discussing relay networks in [11–15] have been proposed to have a source, a relay, and a destination, where transmit DB is employed both at the source and at the relay node. References [12–16] have developed several

DB techniques for relay networks with flat and frequency selective fading channels.

For the WSN maritime communication application, selected nodes collaboratively forward the radio signal to the distant node resulting in the enhancement of the coverage of the WSN. DB is a cooperative scheme that plays a major role in harnessing the limitation in transmit power of the individual sensor devices of the WSNs. The sensor devices are usually battery driven and are deployed in remote areas where periodic battery replacement is unlikely. Therefore, these devices have to rely on their battery power supply for a longer period of time. In DB, wireless sensor nodes minimize the energy spent per node by cooperatively transmitting the same signal simultaneously so that the signal from each node is added constructively at the receiver. Thus, the gain of the distributed array increases with the number of nodes. This further means that transmit signal power per node decreases by the increasing number of nodes for a fixed bit energy-to-noise power spectral density ratio (E_b/N_0) at the receiver.

However, prior to transmission, the beamforming nodes have to share information in a coordinated way, so that all nodes can transmit the same signal. In other words, synchronization is required among the cooperating sensing nodes. Furthermore, the information exchange during the pre-transmission phase creates overhead increase. As a result, the energy consumption of the network increases. In fact, the optimization of energy consumption will be a new research direction for each application of WSNs described in Section 1.2. Particularly for the DB technique for WSNs, it is critical to choose the right number of nodes to be used in beamforming to optimize the total energy consumption of the network and thus to maintain greenness. Therefore, the challenge is how to select the optimum number of nodes and accordingly evaluate the performance of DB for WSNs. In [17, 18], a node selection method based on studies of the differences in beam width of a broadside array and an end-fire array is suggested. The proposed NS method selects those wireless sensor nodes, which are able to form a full size end-fire array and a reduced size broadside array. Further in this direction and as mentioned above, the challenging synchronization requirement and its impact on the system performance are interesting themes for the future research in this field. Moreover, as depicted in Figure 1.4, the vehicular sensor networks [19] for the Intelligent Transportation Systems (ITS) are suggested as another promising and prominent application area for this technique, since distributed vehicles (nodes) on the road can collaboratively forward the signal to the distant vehicle with low sidelobe interference level. The technique further

Figure 1.4   Vehicular sensor network for Intelligent Transportation Systems [19].

improves communications among vehicles and transport infrastructure which ultimately results in higher vehicle safety and efficient traffic management.

1.4 Cognitive Radio for WSNs

Current WSNs operate in the ISM band (see Figure 1.5). This band is shared by many other wireless technologies giving rise to degradation of performance of WSNs due to interference. WSNs can also interfere with other services in this band. The proliferation of WSNs will result in scarcity of spectrum dedicated to wireless sensor communications. CR technology for WSNs improves sensor node communication performance as well as spectral efficiency. It is foreseen that CR will emerge as an active research area for wireless network research in the coming years. Currently, some aspects of CR such as interference detection and avoidance [21], adaptive power control, or adaptive modulation [22] have been introduced and implemented in several radio systems. It is expected that within a time frame of 5 years, more systems with radio cognition capability be operational. Unlike conventional radios in which most of components are implemented in hardware, CR uses software implementations

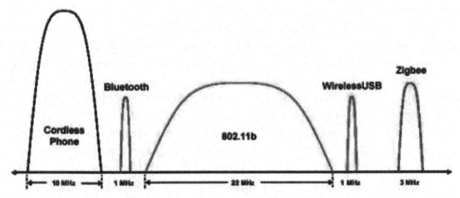

Figure 1.5 Wireless systems operating in the 2.4 GHz ISM band [20].

(i.e., Software Defined Radio[1]) for some functionalities enabling flexible radio operation. The radios are reconfigurable and therefore the need to modify existing hardware is reduced. In this context, the increasing number of sensing nodes equipped with wireless communication capability will require faster connectivity and, thus, wireless spectrum will have to be adapted to the new requirements (of bandwidth). CR will prevent the need to implement hardware upgrades with emergence of new protocols in the future. It will allow the CR-enabled nodes to search for the best frequency-based pre-determined parameters. It should be noted that CR-WSNs differ from conventional WSNs in several aspects. One of the important issues in this regard is interference to other wireless networks or Primary Users (PUs) [24, 25]. Protecting the right of Primary users is the major concern of the CR-WSN. Therefore, miss detection probability of PUs should be minimized in order to minimize interference with the PUs. False alarm probability should also be minimized as large false alarm rates cause spectrum to be under-utilized. High false alarm and miss detection probability in CR-WSNs result in a long waiting delay, frequent channel switching, and significant degradation in throughput. In the context of future WSN research, these issues have to be investigated in detail.

---

[1]Software Defined Radio is a software-based, programmable, and reconfigurable modulation and demodulation technique. With the flexibility that it provides, hybrid platforms can be deployed in the wireless sensor network. By integrating SDR technique in the WSN, with the same (programmable) hardware, more radio standards can be introduced to the network. Therefore, instead of designing again the hardware, only sensor nodes of the WSN are reprogrammed [23].

In particular, applying CR technique to mobile or dynamic WSNs is challenging. One major example of mobile WSNs is the vehicular networks. Adaptive and reconfigurable Vehicle-to-Vehicle (V2V) and Vehicle-to-Infrastructure (V2I) communications have numerous applications (such as emergency warning systems of vehicles, Cooperative adaptive cruise control, Cooperative forward collision warning, intersection collision avoidance, Highway-rail intersection warning, etc.; see Figure 1.6). The flexibility and the agility offered by CR are very useful for resilient communication among mobile wireless nodes (cars), for example, in platooning of automated vehicles of future or in the emergency situations where a CR network should be reconfigured in real time by trading bandwidth maximization or power minimization for more resilience. Major challenges in this regard are Quality of Service guarantee (e.g., low delay and high reliability in high mobility scenarios) and its trade off with bandwidth. Furthermore, the unique features of the mobile WSNs such as mobility of nodes, topology change, and cooperation opportunities among nodes need to be taken into account. Other challenging topic in this direction is dynamic spectrum access for mobile WSNs. In this regard and specifically, the spectral holes for mobile CR-WSNs working in highway scenarios are important as highways (unlike downtown and urban areas) are open spaces and there is a high chance of finding a spectrum hole that can be used opportunistically. This in turn can answer some of bandwidth and congestion problems of the network. These are very interesting and challenging subjects for future research in this specific application area of WSNs.

It is worth mentioning at this point that recently autonomous cars driving on public roads with vehicle-to-infrastructure (V2I) and vehicle-to-vehicle (V2V) communication have been widely demonstrated among others on Dutch public roads in [28]. Thanks to intelligent wireless sensing and networking,

Figure 1.6   V2V (left) and V2I communication (right) [26, 27] respectively.

this technology has potentials to improve road throughput and reduce fuel consumption, while not compromising safety. It is expected that, within a few years, more CR sensing aspects will be implemented in this prototype.

Another interesting and important topic for future research and development of WSNs is the radio channel. In WSN, typically the nodes have low-height antennas. In these applications, the radio propagation channel characteristics and among others the path loss exponent is considerably different from the free-space channel. Therefore, routes with more hops and with shorter hop distances can be more power efficient than those with fewer hops but longer hop distances. To this end, carrying out research on CR-WSN channel and investigating the impact of dynamic channel on the adaptive self-configuration topology mechanism of sensor network is highly suggested to gauge to which extent this can reduce energy consumption and increase network performance.

The research on CR for WSN positions well in the scope of future developments of WSNs as CR will be the enabling technology for WSN that have to coexist with other networks [29] and have heterogeneous devices. The reconfigurability, adaptation, and interoperability capabilities of the CR are major accents of this technology which are pronounced in the research perspectives of future WSNs and are expected to appear in various products by the beginning of the next decade.

1.5 Joint Sensing and Communication in One Technology for WSNs

As is clearly mentioned in the first Conasense book [30] a very important integration strategy concerns communications, positioning, and sensing systems. This integrated vision involves an "active" integration with new business opportunities able to merge three worlds – communications, navigation, and local/remote sensing – that have been apart for years. This vision is the focus of the Communications, Navigation, Sensing, and Services (CNSS) paradigm. In CNSS, communications, navigation, and sensing systems can mutually assist each other by exploiting a bidirectional interaction among them (see Figure 1.7), [31].

Integrated sensing and radio communication systems have emerged in sake of system miniaturization and transceiver unification. With the current technological advancements, the radio frequency front-end architectures in sensing networks and radio communications become more similar. Orthogonal Frequency Division Multiplexing (OFDM) as a capable technology already and

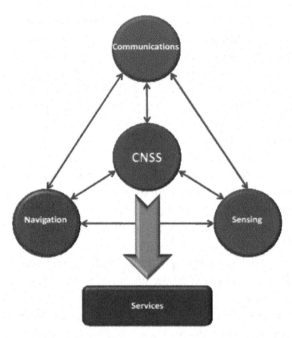

Figure 1.7   Integration of Communications, Navigation, Sensing, and Services [31].

successfully used in wireless communications (e.g., in IEEE802.11a,g,n,p) can be used in WSN. Among the state-of-the art transmission schemes, the seminal concept of [32] on joint ranging (location) and communication using OFDM technology is important. In this context, OFDM can find applications in future WSNs, where context-awareness (location information) is important and, in addition to sensing, data exchange among nodes or between nodes and fusion center is essential. It is worth mentioning at this juncture that:

(i) In addition to reconfigurability and adaptation capability of OFDM, using this technology besides achieving both functionalities simultaneously, the bandwidth will be efficiently used. This theme will be discussed in detail in the next Conasense book.

(ii) In vehicular networks due to Doppler shift, which is a result of high node mobility, the bit error rate performance degrades [33]. By using OFDM, the Doppler effect can be detected in real time, which may be used to correct the highly mobile communication channels for high-speed data link of V2I or V2V of Vehicular Sensing Networks. Another application is a network of UAVs in the air which sense the environment. In this scenario, the UAV nodes are highly mobile and, due to the high-speed regime, the Doppler should be estimated and compensated. OFDM is a capable technology in this regard.

1.6 Physical Layer Security of WSNs

Security is a key issue for the radio sensor networks especially when working in a hostile environment. This has to be considered in the design and development of the whole network. Secure WSN should be designed and implemented and their performance in terms of low latency, reliability, and survivability should be studied. State-of-the art radio transmission schemes (in different layers) have to be developed guaranteeing the required link security. Furthermore, the network of radio sensor nodes should be protected against intrusion. To this end, advanced techniques providing security in WSN should be further studied and should be properly modified according to the requirements of the specific multi-disciplinary application. The challenges in the security and safety of WSNs include: Incorporating security as a design requirement for distributed radio sensor networks, Development of multi-layer radio transmission schemes guaranteeing the required link security, Development of protection schemes against intrusion and spoofing, Study of survivability of WSN, and, for radar sensor networks working in military scenarios, Development of low probability of detection techniques of radio communication links among sensor nodes. As mentioned above, security is one of the most challenging issues in WSN research and development. Because of this importance, this issue has top priority in WSN research and development agenda and is therefore considered to be discussed separately and in detail in the 5th Conasense book. Owing to the wide range of this subject and the space limit of this chapter in the rest of this section, only some major challenges of "Physical Layer Security" are summarized. Generally speaking, computer scientists and engineers have tried hard to continuously come up with improved cryptographic algorithms to address the security issue challenge. But typically, it does not take too long to find an efficient way to crack these algorithms. With the rapid progress of computational devices, the current cryptographic protocols are already becoming more unreliable [34]. A new paradigm which is proposed is the 'Physical Layer Security'. Unlike the traditional cryptographic approach which ignores the effect of the wireless medium, physical layer security exploits the important characteristics of wireless channel (such as fading, interference, and noise) for improving the communication security against eavesdropping attacks. As an example, Figure 1.8 shows the basic idea

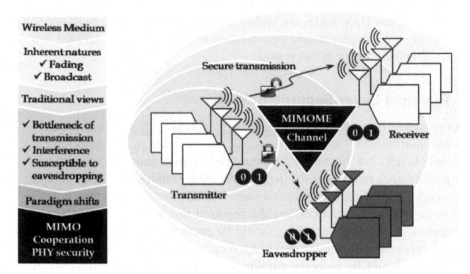

Figure 1.8   MIMO technique for physical-layer secure wireless communications [35].

of MIMO transmission to exploit the properties of physical layer for secure communication [35]. This new security paradigm is expected to complement and significantly increase the overall communication security of future WSN. Further research on the Information Theory and Signal Processing techniques, and approaches for Physical Layer Security of communications in WSNs are envisaged. Methods such as key generation and agreement over wireless channels, various multiantenna (MIMO) transmission schemes, and efficient resource allocation methods in OFDM systems are a few ideas to name.

1.7 Wavelet Technology for Context Aware and Reconfigurable WSNs

As already mentioned in WSNs, several nodes operate at the same frequency band in a network. They share the spectrum, and therefore may interfere with each other. Accordingly, in these networks, the interference becomes a major problem to be addressed. In this direction, and in order to cope with the interference in an intelligent way, research on the context aware design of communication signals for adaptive wireless sensing networks is suggested. On the other hand, the Wavelet packet transform has recently emerged as a novel signal design technique with advantages such as good time–frequency resolution, low sidelobes, and the reconfigurability [36]. The wavelet approach is advantageous for the signal design of smart radar sensor networks [29] mainly because of its flexibility, lower sensitivity to distortion and interference, and better utilization of power and spectrum. Figure 1.9 shows a distributed radiosensor system where several separated transmitter-receiver radio sensors operate in a network and share the spectrum. The radio sensor nodes TX1 and TX2-a,b,c transmit their signal to RX1 and RX2, respectively. The idea is the use of different waveforms by radio sensors concurrently sharing the spectrum such that a particular receiver can separate the resulting received signal and perform its data detection. As radiosensors share the spectrum in order to avoid interference among them, their signals should be orthogonal. This orthogonality is the key characteristic of the distributed radio sensor network. Furthermore, and particularly for radar sensing scenarios, in order to have a high resolution in multiple target detection, the transmitted radar signals should have low correlation sidelobe levels.

In the design, the orthogonal properties of wavelet packet-based basis functions can be well exploited for the situation shown in Figure 1.9. The main advantages of using wavelets are their ability and flexibility to characterize

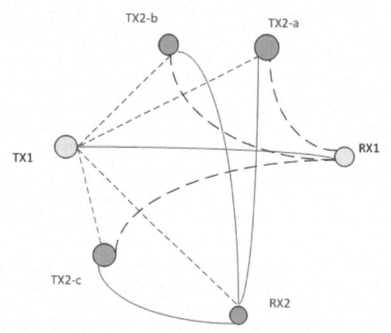

Figure 1.9   Example of shared spectrum distributed radio sensing network configuration [29].

radio signals with adaptive time-frequency resolution. In fact, unlike OFDM modulation that divides the whole bandwidth into orthogonal and overlapping sub-bands of equal bandwidths, Wavelet packet modulation (WPM) assigns wavelet sub-bands having different time and frequency resolutions [36]. Furthermore, WPM is more robust to interference (an extremely important issue when it comes to distributed radio sensor networks where the radio sensing nodes interfere with each other) as well as multipath propagation [36]. Wavelet packet signals are more spectrally efficient than OFDM and, moreover, the advantage of wavelet transform lies in its flexibility to customize and shape the characteristics of the waveforms for joint range sensing and communication functionalities.

For the minimum interference to the adjacent bands, the wavelet signals should be maximally frequency selective. Commonly known wavelets are not frequency selective in nature and hence result in poor spectrum sharing performance. To alleviate this problem, and in the scope of interference problem of WSNs, the design of a family of wavelets that are maximally frequency selective in nature is suggested. To this end, the design constraints should be first enlisted. Then, the challenging problem, most likely non-convex, should be first reformulated into a convex optimization problem and should be solved using Programming tools. In Figure 1.10, results of a newly designed wavelet signal for the context-aware (spectrum sharing) radio sensing networks of Figure 1.9 are shown. It is clear that the designed wavelet provides a superior performance (in terms of interference to adjacent bands) when compared with standard wavelets which are not frequency selective in nature. Details can be found in [29].

It has to be mentioned that, in addition to interference, the wavelet design framework of this research will be easily applied to other design criteria (e.g., low power and greenness of the sensor network, spectrum, throughput or latency performance, timing error or synchronization constraint, or even security of WSNs) by merely altering the objective function of the design procedure. From these, the optimization of power and spectrum are eminently important. However, to be able to do so, the desirable properties of the wavelet bases must be translated into realizable objective functions. This can at times be challenging.

WSN work on the principle of adaptive distributed load sharing among the constituent nodes. They also exploit spatial correlation between the data collected between nodes that are physically close together. Wavelets can also be used to exploit spatial correlation between the data collected between sensing nodes that are physically close together. Accordingly, distributed

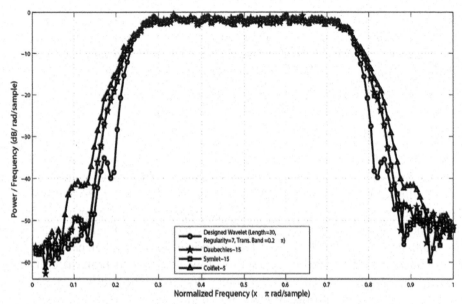

Figure 1.10 Spectrum sharing of radio sensing sources of Figure 1.9 with the newly designed wavelet compared with those based on the standard wavelet family [29].

multi-resolution algorithms are suggested to be developed to reconstruct the information gathered by the nodes with the sensors spending as little energy as possible. In this research direction, wavelets are designed for adaptive distributed processing algorithms in large WSN for power efficient data gathering through use of spatially correlated data. A wireless network architecture is suggested which will efficiently support multi-scale communication and collaboration among sensors for energy and bandwidth efficient communication, while reducing communication overhead, and saving energy.

Furthermore, with regard to the capability of wavelets in signal compression and particularly its importance for the big data challenge of WSNs and in mitigating the burden of the storage utilization as well as in mitigating data congestion, research and development on data compression capability of wavelets used for the future WSNs, in which data are compressed before they are sent out, are suggested. Noticing the proliferation of WSNs in smart cities, homes, factories, cars, Internet-of-Things (IoT), etc., and the imminent big data challenge on one side and the excellent data compression capability of wavelet on the other side, it is expected that this technology will emerge as *the* key technology in the big data hype of the coming period. Moreover, it has to be emphasized that due to the nature of wavelets, the technique will be equally beneficial not only in reducing White noise (i.e., de-noising) but also in mitigation of the wide range of other interferences such as partial discharge, corona, lightning, and interference in smart energy networks (Smart Grid), in particular, or emission of other signals and interferences existing in WSNs working in industrial environments, in general.

1.8 Conclusion

In this chapter, the multi-disciplinary applications of WSNs were discussed and particularly the remarkable enabling technologies for the realization of future intelligent WSNs were focused. Flexible software structures will be needed to reconfigure the signal setting and implement the adaptivity of the designed signal for intelligent WSNs. Advanced technologies to face major challenges of future WSNs such as coexistence with other networks and interference, security, reconfigurability, low power, and high spectral efficiency were addressed. These technologies constitute the future strategic research agenda of the multi-disciplinary applications of WSNs.

References

[1] Nyongesa, F., Djouani, K., Olwal, T., and Hamam, Y. (2015). Doppler shift compensation schemes in VANETs. *Mobile Inform. Syst.* 2015, 11. doi: 10.1155/2015/438159

[2] Chamanian, S., Baghaee, S., Ulusan, H., Zorlu, O., Külah, H., and Uysal-Biyikoglu, E. (2014). Powering-up wireless sensor nodes utilizing rechargeable batteries and an electromagnetic vibration energy harvesting system. *Energies* 7, 6323–6339. doi: 10.3390/en7106323

[3] Wood, A., et al. (2008). Context-aware wireless sensor networks for assisted living and residential monitoring. IEEE Network 22(4), 26–33, Aug.

[4] Keshtgari, M., and Deljoo, A. (2012). A wireless sensor network solution for precision agriculture based on ZigBee Technology. *Wirel. Sensor Network* 4, 25–30 http://dx.doi.org/10.4236/wsn.2012.41004, (http://www.SciRP.org/journal/wsn)

[5] Vellidis, G., Tucker, M., Perry, C., and Kvien, C. A real time smart sensor array for scheduling irrigation: commercialization. www.ugacotton.com/vault/rer/2007/p31.pdf

[6] Lian, X. (2013). "Adaptive and distributed beamforming for cognitive radio," *PhD thesis*, Delft University of Technology.

[7] Lian, X., Nikookar, H., and Ligthart, L. P. (2012). Distributed beamforming with phase-only control for green cognitive radio networks. *EURASIP J. Wirel. Commun. Network 2012*, 1687–1499, Published: 27 February.

[8] Ochiai, H., Mitran, P., Poor, H. V., and Tarokh, V. (2005). Collaborative beamforming for distributed wireless ad hoc sensor networks. *IEEE Trans. Signal Proc.* 53, 4110–4124.

[9] Lun, D., Petropulu, A. P., and Poor, H. V. (2008). A cross-layer approach to collaborative beamforming for wireless Ad Hoc networks. *IEEE Trans. Signal Proc.* 56, 2981–2993.

[10] Zarifi, K., Affes, S., and Ghrayeb, A. (2009). "Distributed beamforming for wireless sensor networks with random node location," in *IEEE International Conference on Acoustics, Speech and Signal Processing, 2009. ICASSP 2009*, pp. 2261–2264.

[11] Mudumbai, R., Barriac, G., and Madhow, U. (2007). On the feasibility of distributed beamforming in wireless networks. *IEEE Trans. Wirel. Commun.* 6, 1754–1763.

[12] Yindi, J., and Jafarkhani, H. (2009). Network beamforming using relays with perfect channel information. *IEEE Trans. Inf. Theory* 55, 2499–2517.

[13] Gan, Z., Kai-Kit, W., Paulraj, A., and Ottersten, B. (2009). Collaborative-relay beamforming with perfect CSI: optimum and distributed implementation. *IEEE Signal Proc. Lett.* 16, 257–260.

[14] Havary-Nassab, V., Shahbazpanahi, S., Grami, A., and Zhi-Quan, L. (2008). Distributed beamforming for relay networks based on second-order statistics of the channel state information. *IEEE Trans. Signal Proc.* 56, 4306–4316.

[15] Fazeli-Dehkordy, S., Shahbazpanahi, S., and Gazor, S. (2009). Multiple peer-to-peer communications using a network of relays. *IEEE Trans. Signal Proc.* 57, 3053–3062.

[16] Haihua, C., Gershman, A. B., and Shahbazpanahi, S. (2010). Filter-and-forward distributed beamforming in relay networks with frequency selective fading. *IEEE Trans. Signal Proc.* 58, 1251–1262.

[17] Nikookar, H. (2013). "Green wireless sensor networks with distributed beamforming and optimal number of sensor nodes." in: *Communication Navigation Sensing and Services*, eds. L. P. Ligthart and R. Prasad, ISBN: 13-9788792982391 (Denmark: River Publishers).

[18] Lian, X., Nikookar, H., and Ligthart, L. P. (2012). Distributed beamforming with phase-only control for green cognitive radio networks. *EURASIP J. Wirel. Commun. Network.* 2012, 1687–1499, Published: 27 February.

[19] Sommer, C. (2010). http://book.car2x.org/Vehicular_Networking_Slides. pdf

[20] http://www.eetimes.com/document.asp?doc_id=1273359

[21] Nikookar, H., and Prasad, R. (2009). *Introduction to Ultra Wideband for Wireless Communications*. Springer, Berlin.

[22] Hanzo, L., Wong, C. H., and Yee, M. S. (2002). *Adaptive Wireless Transceivers*. Wiley, New York.

[23] Li, Y. (2009). *Integrating Software Defined Radio into Wireless Sensor Network*. Thesis, Royal Institute of Technology (KTH).

[24] Budiarjo, I., Nikookar, H., and Ligthart, L. P. (2010). "Modulation techniques for cognitive radio in overlay spectrum sharing environment." in *Cognitive Radio: Terminology, Technology and Techniques*, eds. P. D. McGuire and H. M. Estrada, pp. 73–99, ISBN: 978-1-60876-604-8 (New York: Nova Science Publishers).

[25] Budiarjo, I., Nikookar, H., and Ligthart, L. P. (2008). Cognitive radio modulation techniques. *IEEE Signal Proc. Mag.* 25, 24–34.

[26] http://www.traffictechnologytoday.com/news.php?NewsID=56281

[27] http://www.motortrend.com/news/a-safe-connection-drivers-overwhelmingly-support-communicating-cars-dot-says-208757/

[28] http://davi.connekt.nl/

[29] Nikookar, H. (2015). Signal design for context aware distributed radar sensing networks based on Wavelets. *IEEE J. Selected Topics Signal Proc.* 9.

[30] Ligthart, L. P., and Prasad, R. (eds). (2013). *Communications, Navigation, Sensing and Services* (CONASENSE) book. River Publisher, Denmark.

[31] http://www.conasense.org

[32] Genderen, P. v., and Nikookar, H. (2006). "Radar Network Communication," in *6th International Conference on Communications*, Bucharest, Romania, 313–316, June.

[33] Albarazi, K., Mohammad, U., and Al-holou, N. (2014). Doppler shift impact on vehicular ad-hoc networks. *Can. J. Multimedia Wirel. Netw.* 2, 46–56.

[34] Zhou, X., Song, L., and Zhang, Y. (2013). *Physical Layer Security in Wireless Communications*. CRC Press, Boca Raton.

[35] http://cctlab.khu.ac.kr/sub2-1.html

[36] Nikookar, H. (2013). *Wavelet Radio: Adaptive and Reconfigurable Wireless Systems Based on Wavelets*. Cambridge University Press, Cambridge.

Biographies

Dr. H. Nikookar received his Ph.D. in Electrical Engineering from Delft University of Technology in 1995. In the past, he has led the Radio Advanced Technologies and Systems (RATS) program, and supervised a team of researchers carrying out cutting-edge research in the field of advanced radio transmission. He has received several paper awards at international conferences and symposiums. Dr. Nikookar has published about 150 papers in the peer-reviewed international technical journals and conferences and 12 book chapters, and is the author of two books: *Introduction to Ultra Wideband for Wireless Communications*, Springer, 2009, and *Wavelet Radio*, Cambridge University Press, 2013.

Prof. Dr. ir. L. P. Ligthart was born in Rotterdam, the Netherlands, on September 15, 1946. He received an Engineer's degree (cum laude) and a Doctor of Technology degree from Delft University of Technology. He is Fellow of IET and IEEE and Academician of the Russian Academy of Transport.

He received Honorary Doctorates at MSTUCA in Moscow, Tomsk State University, and MTA Romania.

Since 1988, he held a chair on Microwave transmission, remote sensing, radar and positioning and navigation at Delft University of Technology. He supervised over 50 Ph.Ds.

He founded the International Research Centre for Telecommunications and Radar (IRCTR) at Delft University. He is a founding member of the EuMA, chaired the first EuMW in 1998, and initiated the EuRAD conference in 2004.

Currently, he is emeritus professor of Delft University, guest professor at Universities in Indonesia and China, Chairman of CONASENSE, Member BoG of IEEE-AESS.

His areas of specialization include antennas and propagation, radar and remote sensing, satellite, mobile, and radio communications. He gives various courses on radar, remote sensing, and antennas. He has published over 650 papers, various book chapters, and 5 books.

2

Multi-Disciplinary Applications Requiring Advanced IoT and M2M

Sachin D. Babar[1], Neeli R. Prasad[2], Rasmus H. Nielsen[3], Mahbubul Alam[3] and K. C. Chen[4]

[1]Sinhgad Institute of Technology, Lonavala, Pune, Maharashtra, India
[2]CTIF, Aalborg Univeristy, Aalborg, Denmark
[3]Movimento Groups, Plymouth, MI 48170, USA
[4]National Taiwan University, Taipei City, Taiwan

Abstract

Machine-to-Machine (M2M) and Internet of Things (IoT) communications have attracted considerable attention in research communities and have also started to gain momentum from a commercial perspective where operators are starting to offer services within the domains of fleet management, logistics, home automation, etc. At the same time, the more loosely defined, but broader domain of IoT is picking up as what many are seeing as an evolution of M2M. This chapter investigates the fundamental differences of M2M and IoT by starting out with surveying some of the drivers and moving into an analysis of M2M of today from a technological and business perspective. Finally, multi-disciplinary applications are given.

Keywords: Machine-to-Machine, Internet of Things, Business Value, Technical Challenges, Smart Personal Network, QoL, Smart Retail, Smart Transportation, Smart Energy.

2.1 Introduction

Mobile and IP networks are evolving and converging to provide the ultimate common IT infrastructure for trillions of end points, setting the stage for unprecedented growth opportunities throughout the entire value chain.

Role of ICT for Multi-Disciplinary Applications in 2030, 23–58.

The sheer size and scale of this evolution present significant opportunities for all members of the ecosystem. One of the main networking concepts to drive this is Internet of Things (IoT), and it is becoming increasingly important to gain an understanding of today's market drivers and constraints, the vertical industries poised to gain the most and the vision and strategy to go there. Market strategies and business models as well as specific use cases ideally suited to benefit from IoT are key areas to understand in order to be optimally positioned in the value chain.

The use of the term IoT dates back to 1999 [1–3] and the past decade has seen considerable traction around the term especially from research communities around the world. With the still increasing focus on generating new revenue streams and making current operations more efficient, commercial momentum is picking up with multi-billion-dollar investments spearheaded by technology giants such as General Electrics with the Industrial Internet [4] and Cisco Systems with the Internet of Everything (IoE) [5]. IoE brings business opportunities and innovation through the intelligent interconnection of people, processes, data, and things with value-at-stake being estimated to more than 14 trillion dollars over the next decade. This will be enabled by the emergence of devices, sensors, and things making up the millions and billions of end points that represent real-world objects.

At the same time, Machine-to-Machine (M2M) communication has already seen its first commercial deployments. While IoT is focused on the end points and the representation of physical real-world objects, M2M is connectivity centric and most often specific to licensed cellular access and aims at proving the means to transfer data from remote and often mobile locations in specific vertical solutions in order to improve visibility through simple device monitoring and to some extent integrate with business operations and intelligence.

M2M and IoT are picking up momentum as a consequence of many drivers including:

- Standards: Standardization bodies such as ETSI, ITU, 3GPP, OneM2M, and others are working toward unified architectures and protocols.
- Spectrum: Agility of spectrum is rapidly changing with regulations and spectrum re-farming making more spectra available.
- Cost of connectivity: The average revenue per user (ARPU) is declining and is already at a value less than $5.
- Green initiative: Smart cities and green buildings are high on the agenda in the public sector and are seen as key areas for innovations and improved quality of life (QoL).

- Regulatory: Both Europe and North America are seeing increasing regulatory compliance that is driving the joint standardization work across verticals and sectors.
- Increasing automation: Operational excellence in the public and private sector requires more control of real-world assets.

The above drivers are being supported by market disruptions including the availability and pickup of technology such as mobile, sensors, data analytics, cloud, and consumption models that are increasingly moving from Capital Expenditure (CAPEX) to Operational Expenditure (OPEX) as well as the application and services with the build-out of different ecosystems.

With the objectives of IoT, all objects will be able to exchange information and, if necessary, actively process information according to predefined schemes, which may or may not be deterministic. In such an ambient environment, not only users become ubiquitous but also devices and their context become transparent and ubiquitous. With the miniaturization of devices, an increase in computational power, and reduction in energy consumption, this trend will continue toward IoT [6].

Figure 2.1 shows the house for IoT, which is built from all the components required for communication and connectivity [7]. Communication, data processing, identification, localization, and storage will be the pillars for IoT, which will enable any-to-any and anywhere connectivity. Security, sensor device, and network planning will be the base on which the pillars of IoT will reside. IoT will connect things to users, business, and other things

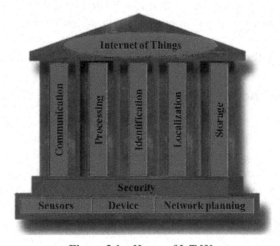

Figure 2.1 House of IoT [3].

using the combination of wired and wireless connectivity. The effectiveness and efficiency of these systems will be important and crucial, which will enable new forms of connectivity, which should be inexpensive with support to standard Internet protocols. Most of the devices in the IoT will be used in two broad areas:

1. Critical Infrastructure: power production/generation/distribution, manufacturing, transportation, etc.
2. Personal infrastructure: personal medical devices, automobiles, home entertainment and device control, retail, etc.

Critical infrastructure represents an attractive target for national and industrial espionage, denial of service, and other disruptive attacks. Internet-connected things that touch very sensitive personal information are high-priority targets for cyber criminals, identity theft, and fraud. Both these areas will demand a new technology requiring new approaches to security and a major change in the way security is architected, delivered, and monitored.

IoT will demand new approaches to security like a secure lightweight operating system, scalable approaches to continuous monitoring and threat mitigation, and new ways of detecting and blocking active threats. Security and privacy aspects are among the most challenging topics in such an interconnected world of miniaturized systems and sensors. Having every 'thing' connected to the global future IoT and communication with each other, new security and privacy problems arise, e.g., confidentiality, authenticity, and integrity of data sensed and exchanged by 'things'. Owing to a manifold of aspects that involves, security for IoT will be a critical concern that must be addressed in order to enable several current and future applications [8, 9].

Figure 2.2 shows the IoT Objectives followed by their description.

2.1.1 Naming and Addressing

Today's Internet addressing scheme is rather rigid; it is well suited to a static, hierarchical topology structure. It provides a very efficient way to label (and find) each device interface in this hierarchy. To support mobility and routing, the next-generation Internet must provide ways to name and route to a much richer set of network elements than just attachment points. A clean architectural separation between name and routable address is a critical requirement for IoT [10, 11].

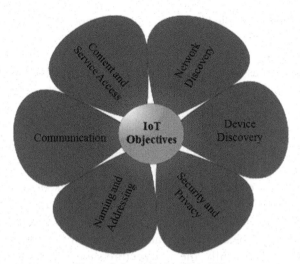

Figure 2.2 IoT objectives [3].

2.1.2 Device Discovery and Network Discovery

The current Internet is text-dominated with relatively efficient search engines for discovering textual resources with manual configuration. An Internet dominated by unstructured information supplied from large numbers of sensor devices must support efficient mechanisms for discovering available sensor resources. The new architecture must support methods for the registration of a new sensor system in the broader network [12, 13].

2.1.3 Content and Service Access

A new architecture should provide data-cleansing mechanisms that prevent corrupted data from propagating through the sensor network. In particular, services that maintain device calibration and monitor/detect adversarial manipulation of sensor devices should be integrated into sensor networks. This could be realized through obtaining context information, metadata, and statistical techniques to locally detect faulty inputs [12, 14, 15].

2.1.4 Communication

Wireless devices should be able to operate independently of the broader Internet. In particular, there may be times during which the connection of a wireless device or, network, to the Internet is not available. During these times, wireless devices should be able to operate stably in modes disconnected from the rest of the infrastructure, as well as be able to opportunistically establish "local" ad-hoc networks using their own native protocols. In particular, this means that issues such as authorization and updating the device state should be seamless, with minimal latency [11, 15].

2.1.5 Security and Privacy

Wireless networks can be expected to be the platform of choice for launching a variety of attacks targeting the new Internet. At the most basic level, wireless devices will likely have evolving naming and addressing schemes and it will be necessary to ensure that the names and addresses that are used are verifiable and authenticated. One parameter uniquely associated with wireless networks is the notion of location. Location information provided by the network should be trustworthy [15]. Additionally, the architecture should provision hooks for future extensions to accommodate legal regulations.

This chapter investigates the evolution from M2M communication to a fully IoT-connected world and which are some of the specific capabilities that will be required in order to facilitate this transformation—a transformation that will result in a brand new networking paradigm toward multi-disciplinary applications.

2.2 Machine-to-Machine: IoT as of Today

The global market for cellular M2M connectivity services is to reach $35B by 2016 [16], resulting in a vast amount of machines and autonomous devices connected to the Internet either via fixed or wireless networks. In many or most cases, wireless connectivity provides the most feasible, effective, and economical link between machines and the corresponding operation or business services. In this manner, M2M combines telecommunication and information technology to automate processes, e.g. to integrate a company's assets with its IT system and to create value-added services. M2M utilizes a wireless data connection as a link between systems, remote devices or locations, and individuals. This link is typically used for collecting information, setting parameters, sending or receiving indications of unusual situations, or taking care of online transactions on countless machines such as elevators, vending machines, automatic metering, etc. As a result, business processes become more efficient and the operational excellence improves. New M2M applications are continuously emerging in a still increasing range of business areas and physical environments. Some of the verticals that have demonstrated potential for concrete business value in M2M are: Telemetry, telematics and in-vehicle applications, public traffic services, industrial applications, security and surveillance, sales and payments, fleet management, telemedicine, and public safety services.

Some of the main barriers for M2M as of today are the complexity of the overall ecosystem along with the lower revenue per unit or device. While the basic functionality in M2M is clear in terms of improving business process efficiency and increasing employee productivity by automating decision and tasks by means of listening, sensing, monitoring, measuring, reporting, and responding, including service delivery from a distance location, the profitability in the value chain is unclear. Looking at market sizing, multiple industry analysts are expecting the global value of M2M (cellular and non-cellular) to exceed $50B by 2015. Communication service providers (CSPs) are still trying to understand and define their role in this new value chain spanning from CSPs positioning themselves as a smart bit-pipe provider to CSPs that want to play a pivotal role in the value chain so that they can tap into new revenue streams beyond connectivity.

For CSPs to be successful in M2M, it is important to understand what the customers' care about and these can be divided into four pillars:

- Financial models:
    - New revenue streams
    - Revenue uplift
    - Reduced cost
- Operating model:
    - Faster response
    - Downtime avoidance
    - Auto replenishment
- Customer model:
    - Better experience
    - Simpler integration
    - More interaction
- New value propositions:
    - New bundled services
    - New revenue models
    - New products.

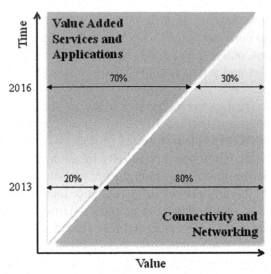

Figure 2.3   The transition of the M2M value chain [2].

Each of the above items relates to one or more specific use cases within a business and M2M is uniquely positioned to support this from the perspective of bringing insight into operations and leveraging the connectivity and the harvesting of data as a way to improve the existing business and generate new.

The M2M market and its value-chain are in a transition phase. In the next 2–3 years, the majority of the revenue allocation will shift from basic connectivity services to more advanced value-added services and applications as shown in Figure 2.3. These advanced services include:

- Asset management of sensors and actuators.
- Big data management for data analytics in the cloud.
- Integration of data intelligence with business intelligence such as customer relationship management (CRM) and enterprise resource planning (ERP).
- Storage of transactional data.
- Compliance management for regulatory requirements.
- Policy manager for process automation.
- Data integrity and security.
- Device identity management.
- Service-level agreement (SLA) report for service availability.
- Activation/deactivation of devices and services on the fly.

- Mobile data plan management.
- Vertical specific systems integrator.

The key questions to answer and successfully address the M2M market are; what is the value of the above services and when will they be crucial for the market?

2.3 Moving to IoT

As presented in the previous section, M2M is in full swing, but with plenty of questions still to be addressed. While IoT is not substantially different from M2M, there are a number of differentiators including the transition from a connectivity and business-centered view to a view centered around devices. At the same time, there is a major gap when considering the functionality and the business value in M2M and IoT, respectively. Figure 2.4 illustrates the evolution of M2M and IoT as seen today toward the future IoT. To summarize, the M2M as seen today is focused around licensed wireless technologies (2G, 3G, and 4G cellular services)and provides functionality targeted at today's vertical use cases to propose the business value. The M2M solutions provided today mostly reside on vertical platforms servicing only one specific vertical. IoT, on the other hand, is much broader when considering connectivity including both licensed and unlicensed wireless along with wired connectivity. Looking at IoT from a business value perspective, it is lacking behind M2M with only few business cases such as the commercialization of RFID tags in supply chain management and logistics. While there are some IoT platforms already out there today, these are more addressing the market from an M2M perspective. The IoT to be seen tomorrow is a concept that moves beyond the basic connectivity and technological innovations and merges the

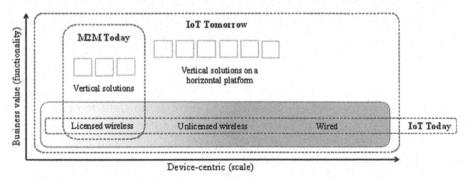

Figure 2.4   The evolution from M2M and IoT as of today to IoT of tomorrow [2].

gap to the envisioned use cases in order to bring the needed functionality and business values. This includes a focus on platforms and how solutions are delivered with horizontal platforms that are able to support a multitude of vertical solutions. There are both business and technological challenges involved in moving toward IoT and some of the main technological challenges concern paradigms for networking and computing as outlines in the below subsections.

2.3.1 Networking

The things in IoT cover a large range of devices with different connectivities and levels of intelligence from full-fledged industrial PCs or gateway devices over small software stacks residing on the assets to passive RFID tags that are only brought to live depending on the location and are fully depending on an IT infrastructure. The range of devices alone indicates the challenge in terms of scalability and how this is addressed throughout the network. Emerging protocols including IPv6 over Low power Wireless Personal Area Networks (6LoWPAN) [17] and IPv6 Routing Protocol for Low-Power and Lossy Networks (RPL) [18] along with, for example, the Constrained Application Protocol (CoAP) [19] are suitable for resource-constrained devices.

With the adoption of IoT in manufacturing settings, traditional IP networking with the QoS guarantees provided will no longer suffice compared with proprietary industrial networking solutions and protocols. As a consequence, deterministic networking is becoming still more relevant and is a major focus area in standardization bodies, e.g. in IEEE. Not only deterministic networking, but also QoS and networking performance in general will need to be considered with the introduction of IoT. Most M2M and IoT solutions are currently over the top, meaning that there is no integration with the network. This becomes a challenge when moving toward mission critical use cases such as connected vehicles and telemedicine, where latency and the guarantee of delivery are essential. Therefore, the need exists to consider networking not just as any transport medium, but also as an integral part of the end-to-end solution.

Figure 2.5 shows the high-level view of moving from a networking paradigm where connectivity is assumed and there is no tight integration with the network to a scenario where the network is a part of the ecosystem and provides the functionality needed to support the individual use cases. The functionality spans all the way from guaranteeing the QoS to providing

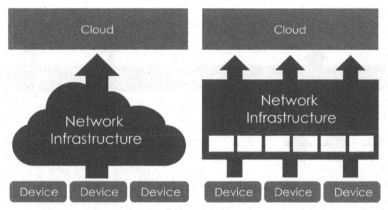

Figure 2.5 Integrating with the network infrastructure [2].

scalable AAA services. In addition, the network will move away from being a passive infrastructure only concerned with transportation into a mesh of networking, computation, and storage units that can take an active part in providing functionality that would traditionally reside on the application layer and only be processed in either end point. Such functionality includes data filtering and aggregation along with analytics and other tasks for which the networking elements are particularly suited including store-and-forward and content management and distribution

2.3.2 Computing

Computing is a main aspect of IoT, where vast amounts of data are available to improve operational and business processes as well as to enhance consumer experience. The computing paradigms of today are focused around cloud computing that has become the computational location of preference due to the commoditization of hardware and the flexibility of resources enabled through the use of virtualization. The cloud will still play a pivotal role in IoT, but will also pose certain limitations relating to especially network capacity and QoS requirements. The forward-looking consensus suggests that IoT will span a new interest in distributed computing, where Figure 2.6 shows an example of the difference between bringing data to the analytics and computation compared with bringing the analytics and computation to the data. Tradeoffs will have to be made depending on the scenario and specific use case. As suggested in the figure, one of the advantages in not bringing everything into

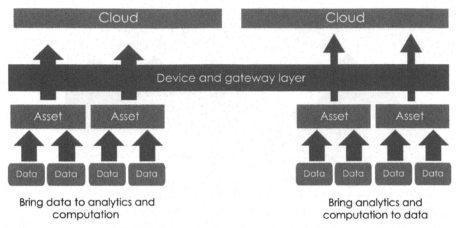

Figure 2.6   Bringing computing to data [2].

the cloud for processing is the reduced load of the network by only sending essential data that have already been pre-processed at the asset or in the device and gateway layer providing the connectivity. This can tremendously save the bandwidth need for harvesting the full value of the data generated by billions of devices, which is advantageous from an economical perspective when considering that a considerable amount of connectivity will be provided over cellular connections powered by bought-and-paid-for data plans.

Saving bandwidth, however, is not the only advantage of processing data closer to the location where it is being generated. The QoS again becomes a relevant means of comparison when considering that cloud resources are primarily utilized due to economical reasons facilitated by having few shared large computing facilities. Just considering the geographical distribution of the underlying data centers suggests that the propagation delay combined with processing and queuing delays had already been setting a lower bound on the latency that can be guaranteed and, in turn, the use cases that can be supported. To add to this, cloud-computing resources are most often shared between customers and services with a large degree of overbooking, making the deterministic aspects hard to evaluate and impossible to guarantee.

2.4 Multi-Disciplinary Applications

Nanotechnology and energy scavenging technologies are packing processing power in smaller space, so that networked computing can become part of the things. These technologies will create an infrastructure, with capabilities organized into global systems. This will serve the information and decision-making needs in adaptive and dynamic ways.

The key feature of the Internet will be innovation in services relying on information related to the identity, status, and possibly changing location of things. IoT brings disruptive business models, and new societal services that will improve the quality of life (QoL) (see Figure 2.8). The future network into which it will one day evolve—will have to deal with an increase in traffic as today's offline things are brought online. This will make industrial processes efficient with higher degrees of productivity.

As the technologies needed for the IoT become available, a wide range of applications and use cases will be developed. These can support policy in areas including transportation, environment, energy efficiency, and health.

Connecting Everything (see Figure 2.7). Innovation is quickly moving from the smart phones in the hands, to the smart devices connecting the world. These devices are taking everyday things and making them infinitely better. Already, there are vehicle accident-reporting devices, mobile payment systems, remote health monitoring devices, smart utility meters, connected refrigerators, picture frames, pill bottle caps, traffic lights, and smart parking

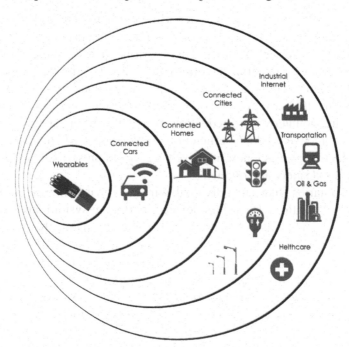

Figure 2.7 IoT Applications: everything connected.

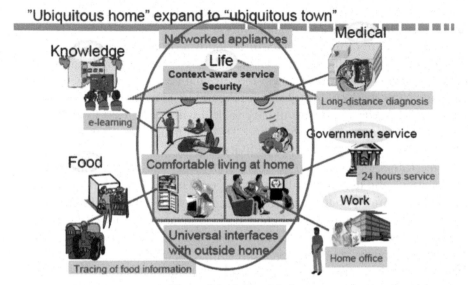

Figure 2.8 IoT and QoL.

meters that use mobile technology to streamline our lives. These devices are poised to transform virtually every sector, from transportation to health care, from energy to consumer electronics. There will be growth in productivity, increasing efficiency in material handling and general logistics, warehousing, product tracking, data management, reducing production and handling costs, speeding the flow of assets, anti-theft and faster recovery time of stolen items, addressing counterfeiting, reduction in manufacture mistakes, immediate recall of defective products, efficient recycling and waste management, CO_2 reductions, energy efficiency, improved security of prescribed medicines, and improved food safety and quality.

The IoT scenarios, like an individual wireless device interfacing with the Internet, constellation of wireless devices, pervasive system, and sensor network, are associated with new network service requirements that motivate rethinking of several Internet architecture issues. Several mobile/wireless features may require mechanisms that cannot be implemented through the conventional IP framework for the Internet, or if they can, may suffer from performance degradation due to the additional overhead associated with network protocols that were originally designed for static infrastructure computing [20].

Few of the potential applications of IoT are described below. Behind the simple, visible functionalities illustrated in these examples, lies an invisible but complex web of networked connections and smart systems, which captures information, processes it, transmits it, and stores it.

2.4.1 Smart Personal Network (PN)

PNs comprise potentially *"all of a person's devices capable of network connection whether in his or her wireless vicinity, at home or in the office"*. This requires major extensions of the present Personal Area Networking (PAN). PNs are configured in an ad hoc fashion, as the opportunity and the demand arise to support a person's private and professional applications. These applications may run on user's personal devices, but also on foreign devices. PNs consist of communicating clusters of devices, possibly shared with others, and connected through various suitable communication means. This is illustrated in Figure 2.9. Unlike PANs, with a limited geographically coverage, PNs have an unrestricted geographical span, and may incorporate devices into the personal environment regardless of their geographic location. In order to extend their reach, they need the support of infrastructure-based networks. The cooperation between PNs belonging to different people in a federation is illustrated in Figure 2.10.

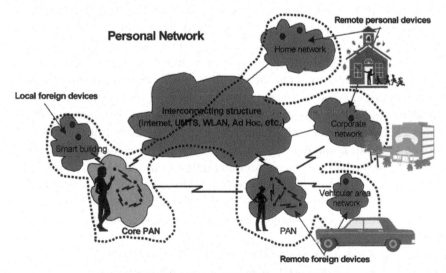

Figure 2.9 Illustration of the PN concept [21, 22].

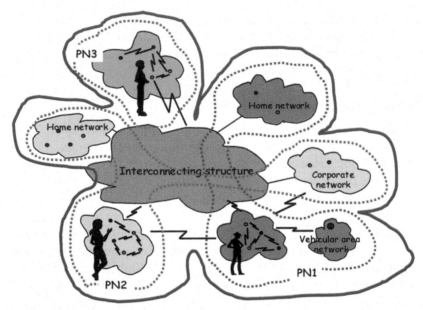

Figure 2.10 Cooperating PNs belonging to different people [21, 23].

2.4.2 Smart Retail

Consider a virtual shopping scenario as shown in Figure 2.11. Suppose, you are at your office, and one of your family members demands for a matching sofa set for your hall. Owing to office constraints, you cannot go to the shopping mall to do the needful. You also do not know about the size and color that will best suit your hall. Now to avoid the travelling back home and going to the shop, you can just call your home network through your mobile device sitting at your office and connect to your home network through different wireless technologies. The home network consists of multiple sensors/wireless devices. You can call in your home network and connect to the camera located in the home. You view the hall and take a remote picture of the hall from a suitable angle. On similar lines, you can connect to the network of the shopping mall, and select the item that best suits your hall. After finalizing the item, now you do the payment by connecting to the bank and transfer the amount to the shopping mall store account.

By using different networks and devices as shown in Figure 2.12, we have just left our homes, mobile, and bank information open to hackers and thieves. Apart from the security present in the existing networks, one will have to focus on the security aspects of all the resource-constrained devices involved in the

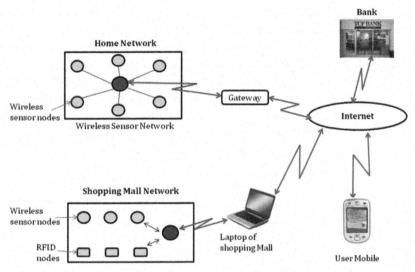

Figure 2.11   Virtual shopping application.

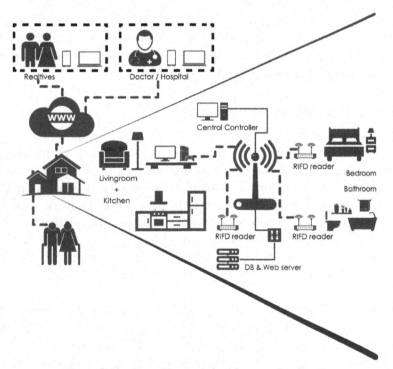

Figure 2.12   Easy life: smart healthcare and wellness.

communications. Existing networks are inadequate to meet the security needs of data-sensitive applications. Hence, in security terms, we need to identify two areas, which need to be secured, i.e. network security and device security.

IoT will fully automate the warehouses, where items will be checked in and out and orders passed on automatically to the suppliers. This will improve asset management and make supply chain management more efficient and proactive [24]. Production and transportation can be adjusted dynamically, saving time and energy and being more environmentally friendly.

The following use cases can be visualized:

- Supply chain management
- Logistics
- Smart shopping list
- Smart shopping/item guidance
- Personalized advertisement
- Payment
- Point-of-sale
- Vending machines and telemetry.

IoT develops an improved ecosystem that connects physical and virtual worlds, allowing bidirectional, real-time interaction with both in- and outside of the store. Customer Experience, supply chain operation, decrease in theft and new channels, and revenue streams will increase.

2.4.3 Smart Healthcare and Wellness

Smart Healthcare and Wellness (see Figure 2.13) is an illustration covering, e.g. the user's interaction with the public systems in terms of the broad area of managing illness, accidents, and general health care perspectives. The case covers the situations in which healthcare professionals in the primary and secondary sector and patients in their home, at work, or travelling communicate between heterogeneous networks establishing virtual clinics. This communication has the potential of providing higher quality and less costly treatment and care, and at the same time, the patients will feel better monitored and have a better feedback on their disease and treatment.

Traditional PCs, tablets, and smart phones as patients already have been combined with medical devices and other gadgets in their body or at disposal when hospitalized. Healthcare professionals at specialized hospitals as well as the local hospitals closer to the patient's home, the patient's general practitioner, and homecare employees each have their own communication

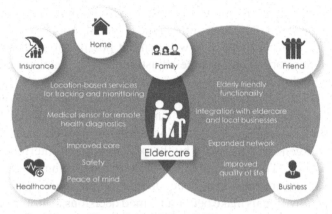

Figure 2.13 IoT and eldercare.

platform and devices through which they need to communicate with each other and other patients.

The possible use cases of this application are:

- Healthcare records
- Healthcare vitals
- Rehabilitation
- Independent living
- Pharmaceuticals
- Insurance.

The Impact of IoT on Healthcare and Wellness will be very high. It will revitalize (see Figure 2.13):

- Better and faster treatment
- Shorter and better recovery
- Improved quality of life
- Longer stay in own home
- Safety
- Cost savings
- Maintaining healthcare plans
- Keeping citizens fit
- More efficient treatment
- Less hospital beds and places in elder centers
- Less hospitalization
- Cost savings in reduced consultations and shorter use of recovery equipment.

Closely linked to this case are relevant issues of IoT communication, security of handling personalized data, trust and seamless global mobility.

Using the IoT concept to the area will make it possible to realize instant and continuous monitoring, and remote adjustments and treatment are some of the prospects in this case.

The use of RFID and sensing technologies as IoT enabler will allow real-time monitoring of patients, leading to earlier diagnosis. Vital parameters such as heart rate, breathing rate, and blood pressure can be measured by lightweight, intelligent sensors, worn by the patient without interfering with their daily activities. These wearable networked monitoring systems can acquire, process, and transmit data concerning multiple health parameters, letting medical professionals make better informed decisions. Automatic alerts can be immediately sent to medical staff to warn of any deterioration in patients' condition [25].

A medication example is the so-called "Smart networked device", which could help to address the challenges of our aging society (see Figure 2.12). For instance, using tags on pharmaceutical products could allow monitoring to check that an aging patient takes the right mix of drugs. Networked sensors could be used to help and monitor the behavior of people with special needs; e.g., sensors might automatically select the right shower temperature, thus avoiding the risk of scalding, or could report unexpected behavior that might require medical intervention.

2.4.4 Smart Transportation

Transport infrastructure—roadways, railways, and runways—were often not designed for the 21st century or to efficiently handle the growing demands. The sector is continuously working on improving the ease of commuting, safety, time and cost savings, efficiency, management, and maintenance. IoT will bring improvement and transformation in use cases below (see Figures 2.14 and 2.15).

- Fare and toll collection, road pricing
- Public transportation
- Cargo and luggage
- Intelligent transport system (ITS)
- Manufacturing and maintenance
- Vehicle to Vehicle (V2V), Vehicle to Infrastructure (V2I)
- Progressive insurance
- Fleet management.

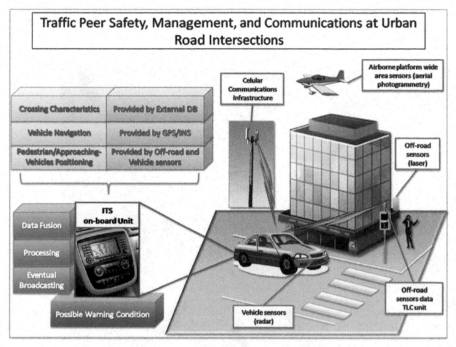

Figure 2.14 Traffic management.

The combination of smart infrastructure and smart vehicles will improve everyday lives by making autonomous cars, accidents less frequent, ease of parking, crashes more survivable, dynamic delivery, and efficient commutes. This will provide

- Ease of commuting: Connected onboard navigation systems that forewarn of traffic back-ups, and getting to destination faster and easier.
- Safety: Vehicle-theft-tracking services that can help track the car if stolen, and shut it down if needed.
- Time and cost savings.
- Management and Maintenance: Onboard engine diagnostic systems that continuously monitor oil pressure, tire pressure, and electronics for preventative maintenance.
- Efficiency: Telematics systems that can connect the car to roadside assistance for a flat tire, empty gas tank, or to automatically call for emergency help in the event of an accident.

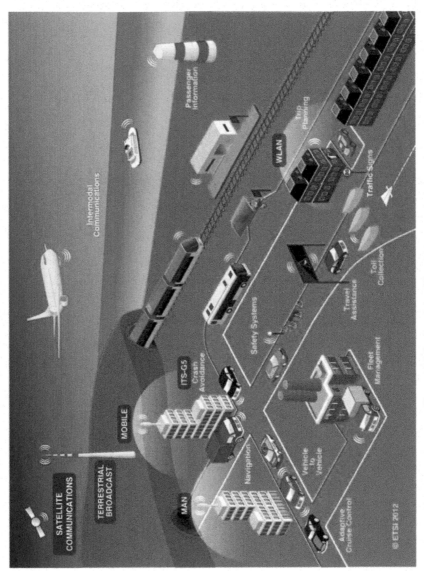

Figure 2.15   Smart intelligent transport system.

Source: ETSI 2012.

2.4.5 Smart Energy and a Better Environment

Over the course of the 20th century, the electrical power systems of industrialized economies have become one of the most complex systems. To realize the full potential of Smart Energy, major changes in the energy system, especially at the lower network level, will require integration of smart houses in the smart grid. Introduction of renewable and the trend toward an all-electric infrastructure lead to an increase in complexity, system management effort, and cost when the smart grid should provide an answer for Smart Houses interacting with Smart Grids to achieve next-generation energy efficiency and sustainability (see Figure 2.16) [26, 27].

- *Bi-directional information exchange*: in house; house-to-grid; grid-to-enterprise; house-to-enterprise: *local and remote communication.*
- *Multi-agent and service-oriented architectures*: Distributed control and information processing.

The use of networked temperature and lighting sensors will allow intelligent houses to reduce energy consumption without compromising comfort, by dynamically adjusting room temperature and lighting conditions. Remote monitoring of home or office energy consumption will allow for better planning of energy needs. The IoT will have a major effect on the way traffic, weather, and particles in the air, water pollution, and the environment can be monitored and statistics collected. Wastewater treatment plants will evolve into bio-refineries with the new, innovative wastewater treatment processes.

For better and green environment, IoT will bring the following:

- Energy savings
- Green conscience
- Optimized use-driven production
- Efficiency
- Carbon emissions.

Control and monitoring is key in such a scenario. Currently, control and monitoring relies on proprietary protocols, that lack the ability to communicate and interoperate. Additionally, within Smart House/Smart Grid, it is recognized that a service-oriented approach in conjunction with an event-based infrastructure is the way to go. Information generated at the point of action (device level) is used by devices at higher-level systems and is aggregated and processed, as well as used by global services.

The smart devices are enabling people to use energy more efficiently in the electric grid, in buildings, in homes, on the farm, in factories, and throughout

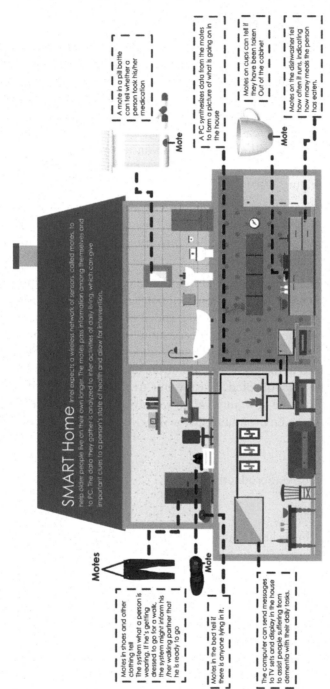

Figure 2.16   Smart home; house of the future.

Source: IEEE Spectrum, Dec. 2004.

environment. Smart grid technology is helping businesses and consumers cut their energy usage while advancing an intelligent, resilient, and self-balancing utility network by enabling utilities to wirelessly connect to circuit breakers and meters. Real-time data from smart meters are also helping utilities monitor use and tackle energy efficiency (see Figure 2.17).

2.4.6 Smart Manufacturing and Logistics

Smart Manufacturing technologies will revolutionize supply chain operations by

- Suspected unapproved parts
- Part traceability
- On-condition monitoring
- Recycling
- Waste management
- Just-in-time production
- High risk production
- Agriculture and farm animals
- Food chain management.

The IoT intelligent systems enable rapid manufacturing of new products, dynamic response to product demands, and real-time optimization of manufacturing production and supply chain networks, by networking machinery, sensors, and control systems together. Digital control systems to automate process controls, operator tools, and service information systems to optimize plant safety and security are within the purview of the IoT. But it also extends itself to asset management via predictive maintenance, statistical evaluation, and measurements to maximize reliability.

Logistics companies were among the first to adopt wireless devices as a means to manage and monitor their processes. Initially, the hand-held devices that delivery drivers used delivered benefits primarily by simplifying and automating existing paper-based processes [28].

IoT allows logistics companies to move beyond simply by making existing processes better by use of low-cost, connected, location-aware devices. This makes it possible to dynamically track both vehicles and the packages [29, 30].

The logistic infrastructure using IoT (see Figure 2.18):

- Vehicles: with modern trucks, planes, locomotives, and ships bristling with sensors, embedded processors, and wireless connectivity.

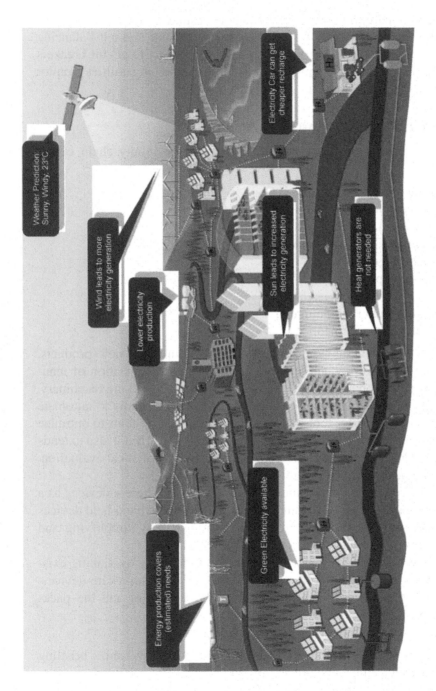

Figure 2.17 Technology platform smart grids.

Source: M. SANCHEZ, 2006, European Technology Platform Smart Grids.

Figure 2.18   The factory floor.

Source: Libelium World [31].

- Sites: Vehicles and containers pass through or park in many sites, e.g. ports, yards, warehouses, and distribution centers. Intelligence and sensing capability can be embedded in the equipment and structures of the sites, including:

  - Mobile equipment, e.g. forklifts, yard tractors, container handlers, mobile cranes, etc.;
  - Stationary or semi-stationary equipment, e.g. cranes, conveyor belts, carousels, automated storage, and retrieval systems;
  - Site structures and ingress/egress points, e.g. sensors in dock doors, yard entry/exit gates, light poles, embedded in floors or under pavement, attached to ceilings and other structures.

- Roads: Intelligence on the roadways, railways, runways, canals/locks, etc.

IoT can address compliance, regulations, and quality reporting requirements such as parts traceability and product genealogy, emissions and country of origin. With IoT, organizations are better suited to trackshipped products for warranties, returns and predictive support for maintenance.

The real premise of IoT-enabled supply chains is to delegate decision making on some of the operational aspects to smart objects and systems, based on real time analytics and machine learning algorithms.

2.5 Conclusions

This chapter presented some of the considerations to be made when moving from M2M into the commercialization of IoT as it is envisioned. The current state of M2M was presented along with drivers and some of the open questions still to be addressed. Some of the challenges with respect to technology when moving into IoT were presented both for networking and computing. IoT and M2M are not well-defined, pick your own definition. Standardization work is still going on. There are many Use cases and applications. IoT will be part of our daily life and will improve QoL.

Technology areas such as WSNs and RFID are enablers of the ubiquitous transmission of data. IoT and M2M utilize technologies to create a world where data can be collected from anywhere and information can be extracted to gain knowledge.

While the IoT is not a new term, the maturity of the industry and underlying technologies are enabling the full-scale commercialization driven by the market spearheaded by the need for operational excellence and improved business intelligence along with improved customer experience.

IoT can help companies and factories provide improved predictability of customer demand with real-time visibility of product and service demand signals. In a supply chain, strategic deployment of IoT technologies can improve asset utilization, customer service, working capital deployment, waste reduction, and sustainability. Real-time communication between machines, factories, logistic providers, and suppliers provides improved visibility on the end-to-end supply chain.

References

[1] Fleisch, E. (2010). "What is the Internet of Things?—An Economic Perspective", *Auto-ID Labs White Paper* WP-BIZAPP-053.
[2] Alam, M., Nielsen, R. H., and Prasad, N. R. (2013). "The evolution of M2M into IoT, Communications and Networking," in: *IEEE First International Black Sea Conference (IEEE BlackSeaCom)*, 112–115. doi: 10.1109/BlackSea-Com.2013.6623392

[3] Babar, S. D. (2015). *Security framework and jamming detection for internet of things*. Ph.D. Thesis.

[4] Evans, P. C., and Annunziata, M. (2012). *Industrial* Internet *Pushing the Boundaries of Minds and Machines*. GE White Paper.

[5] Cisco Systems. What is the Internet of Everything? Available at: http://www.cisco.com/web/tomorrow-starts-here/ioe/index.html

[6] Agrawal, S., and Das, M. L., "Internet of Things—A paradigm shift of future Internet applications", in *International Conference on Engineering (NUiCONE)*, 2011 Nirma University, IEEE, 1, 7.

[7] Babar, S. D. (2015). Security framework and jamming detection for internet of things. Ph.D. Thesis, February 25.

[8] Silverajan, B. and Harju, J., "Developing network software and communications protocols towards the internet of things", in *Proceedings of the Fourth international ICST Conference on Communication System Software and middleware* (Dublin, Ireland, June 16–19, 2009). COMSWARE '09. ACM, New York, NY, 2009, 1–8.

[9] Wang, C., Daneshmand, M., Dohler, M., Mao, X., Hu, R. Q., and Wang, H. (2013). Guest Editorial – Special Issue on Internet of Things (IoT): Architecture, Protocols and Services. *Sensors J. IEEE* 13, 3505–3510.

[10] Seskar, I., Nagaraja, K., Nelson, S., and Raychaudhuri, D. (2011). MobilityFirst future internet architecture project," in *Proceedings of the 7th Asian Internet Engineering Conference AINTEC 2011*, ACM, New York, NY, USA.

[11] Adjie-Winoto, W., Schwartz, E., Balakrishnan, H., and Lilley J. (1999). "The design and implementation of an intentional naming system," in: *Proceedings of the Seventeenth ACM Symposium on Operating Systems Principle, SOSP '99*. ACM, New York, NY, 186–201.

[12] Beerliova, Z., Eberhard, F., Erlebach, T., Hall, A., Hoffmann, M., Mihal'ak, M., and Ram, L. S. (2006). Network discovery and verification. *IEEE J. Selected Areas Commun.* 24, 2168–2181.

[13] Jara, A. J., Lopez, P., Fernandez, D., Castillo, J. F., Zamora, M. A., and Skarmeta, A. F. (2014). Mobile digcovery: discovering and interacting with the world through the Internet of things. *ACM Pers. Ubiquitous Comput.* 18, 2, 323–338.

[14] Wei, Q., and Jin, Z. (2012). "Service discovery for internet of things: A context-awareness perspective", in *Proceedings of the Fourth Asia-Pacific Symposium on Internetware (Internetware 2012)*, ACM, New York, NY, USA, Article 25, 6.

[15] Hu, Y.-C., and Wang, H. J. (2005). "Location privacy in wireless networks", in: *Proceedings of the ACM SIGCOMM Asia Workshop*, 1–5.

[16] ABI research, "Cellular Machine-to-Machine (M2M) Markets", 2012.

[17] Yibo, C., Hou, K. M., Zhou, H., Shi, H. l., Liu, X., Diao, X., Ding, H., Li, J. J., and de Vaulx, C. (2011). "6LoWPAN Stacks: A Survey," in *7th International Conference on Wireless Communications, Networking and Mobile Computing (WiCOM)*, 23–25 Sept., ISSN: 2161-9646.

[18] Winter, T., Thubert, P., Brandt, A., Hui, J., Kelsey, R., Levis, P., Pister, K., Struik, R., and Vasseur, JP. (2012). "RPL: IPv6 Routing Protocol for Low-Power and Lossy Networks," in *Internet Engineering Task Force (IETF), Request for Comments: 6550, Category: Standards Track*, ISSN: 2070-1721.

[19] Shelby, Z., Hartke, K., and Bormann, C. (2014). "The Constrained Application Protocol (CoAP)," Internet Engineering Task Force (IETF), Request for Comments: 7252, Category: Standards Track, ISSN: 2070-1721.

[20] Wang, C., Daneshmand, M., Dohler, M., Mao, X., Hu, R. Q., Wang, H. (2013). Guest editorial–special issue on Internet of Things (IoT): architecture, protocols and services. *Sensors J. IEEE* 13, 3505–3510.

[21] MAGNET Beyond; IST 027396 MAGNET Beyond: Making Personal Networks Happen.

[22] Niemegeers, I. G., and Heemstra de Groot, S. M. (2002). "From personal area networks to personal networks: a user oriented approach", in *Special Issue of the journal Wireless Personal Communication*. Kluwer, Netherlands.

[23] Niemegeers, I. G., and Heemstra de Groot, S. M. (2003). Research Issues in Ad-Hoc Distributed Personal Networking. *J. Wirel. Pers. Commun.* 26, 149–167.

[24] Labs, S. R. L. (n.d.). *Future Retail Center, SAP Research Living Labs*. Available at: http://www.sap.com/corporate-en/ourcompany/inno vation/research/livinglabs/futureretail/index.epx

[25] Nussbaum, G. (2006). "People with disabilities: assistive homes and environments," in *Computers Helping People with Special Needs*.

[26] Li, X., Lu, R. X., Liang, X. H., Shen, X. M., Chen, J. M., and Lin, X. D. (2011). Smart community: an Internet of Things application. *IEEE Commun. Mag.* 49, 68–75.

[27] Vermesan, O., and Friess, P., (eds.) (2103). *Internet of Things: Converging Technologies for Smart Environments and Integrated Ecosystems*. River Publishers, ISBN: 978-87-92982-73-5.

[28] http://cerasis.com/transportation-technology/mobile-tms-application/

[29] Zhang, Y., and Sun, S. (2013). "Real-time data driven monitoring and optimization method for IoT—based sensible production process, networking, sensing and control (ICNSC)," in 2013 10th IEEE International Conference, 486–490. doi: 10.1109/ ICNSC.2013.6548787

[30] Yuqiang, C., Jianlan, G., and Xuanzi, H. (2010) "E research of Internet of things' supporting technologies which face the logistics industry," in Computational Intelligence and Security (CIS) International Conference, 659–663, doi: 10.1109/CIS.2010.148

[31] http://www.libelium.com/smart-factory-reducing-maintenance-costs-en suring-quality-manufacturing-process/

Biographies

Dr. S. D. Babar is Associate Professor and Head of Department of Computer Engineering, SIT Lonavala, India. He is ISTE Life Member. He is awarded with the Phd Degree on February 25, 2015 from Aalborg University, Denmark, in the area of Wireless communication. He did B.SC. and M.Sc. in Computer Engineering from Savitribai Phule Pune University, Pune, Maharashtra, India, in 2002 and 2006, respectively. He has 14 years of teaching experience. He has published more than 30 papers at national and international level. He has authored two books on subjects like Software Engineering and Analysis of Algorithm & Design. He has received the Cambridge International Certificate for Teachers and Trainers at Professional level under MISSION10X Program. He is IBM DB2-certified professional. His research interests are Data Structures, Algorithms, Theory of Computer Science, IoT, and Security.

Dr. N. R. Prasad is a security and wireless technology strategist, who through her career has been driving business and technology innovation, from incubation to prototyping to validation. She has focus and the abilities to transform organizations and networking technologies to address changes in markets. She has made her way up the waves of secure communication technology by contributing to the most groundbreaking and commercial inventions. She has general management, leadership, and technology skills, having worked for service providers and technology companies in various key leadership roles.

She is leading a global team of 20+ researchers across multiple technical areas and projects in Japan, India, throughout Europe and USA. She has been involved in projects and plays a key role from concept to implementation to standardization. Her strong commitment to operational excellence, innovative approach to business and technological problems, and aptitude for partnering cross-functionally across the industry have reshaped and elevated her role as project coordinator making her the preferred partner in multinational and European Commission project consortium.

Her notable accomplishments include enhancing the technology of multinationals including CISCO, HUAWEI, NIKSUN, Nokia-Siemens, and NICT, defining the reference framework for Future Internet Assembly and being one of the early key contributors to Internet of Things. She is also expert member of governmental working groups and cross-continental forums.

Previously, she has served as chief system/network architect on large-scale projects from both the network operator and vendor looking across the entire product and solution portfolio covering security, wireless, mobility, Internet of Things, Machine-to-Machine, eHealth, smart cities, and cloud technologies. She was one of the key contributors to the commercialization of WLAN for which she has published two books.

Dr. R. H. Nielsen is Director of product management and marketing at Movimento for over-the-air software updates in automotive and has more than ten years of R&D and business experience in industry and academia. He is the founder and partner of R&D intensive network planning companies and is working in the field of Communications and Networks focusing on Secure Network Architectures, Wireless Sensor Networks, Internet of Things, Optical Network Planning and Optimization, Virtualization, and Management. He served from 2010 in academia as Postdoc and from 2011 to 2012 as Assistant Professor. He is the Technical manager in several industrial and European Commission (EC) projects and has involved in project management and write-up for several project proposals. Formerly, he was principal architect in New Ventures as part of IOTG, Cisco Systems, driving IoT platform, edge computing, and deterministic Ethernet with broad R&D knowledge in architectures, operational research, and optimization, and expertise in applied R&D for market and industrial needs. His current focus is on Internet of Things (Wireless Sensor Networks, RFID, etc.), planning, monitoring, optimization, security and virtualization of next generation networks/future internet (NGN/FI) and architectures.

M. Alam, CTO/CMO Movimento Group, adding deep entrepreneurial skills to international technology management experience and fluency in four languages, Mahbubul Alam is a change driver as CTO/CMO of Movimento.

A frequent author, speaker, and multiple patent holder Alam was brought to the company in early 2015 to reinvent technology and strategy, leading a transformative era in which Movimento will help shepherd the auto industry through the biggest changes since Henry Ford's days. Prior to joining Movimento, Alam spent 14 productive years as a groundbreaking technologist and strategist at Cisco on two continents. He began in the mobile technology arena, later working in a variety of business development and market intelligence capacities in Cisco's Netherlands operation. Having developed a solid name in the company for far-sighted, winning strategies, he was sent to Silicon Valley to head up Cisco's Internet-of-Things (IoT) and Machine-to-Machine (M2M) platforms in 2012.

After coming to California, he proceeded to create a series of now-renowned products and businesses at Cisco: The flagship 4G LTE multi-service router, the IoT edge-cloud gateway, the next-generation Integrated Services Router (ISR), Enterprise Mobility Solutions and more. Alam grew the company's M2M business from nothing to $350 million in four years. Along the way, he also helped initiate the company's smart connected car roadmap. However, Alam also created several successful internal startups within Cisco, such as IoT vertical solutions for the transportation, manufacturing, and mining industries. He drove corporate investment decisions, ranging between $100 million and $3 billion and directed numerous Cisco portfolios, including the $4 billion Access Routing Technology Group.

Although Alam had created and built a small chain of top-rated dining establishments in the Netherlands that began while he was in college, he moved into the technology arena when he was still in his mid-20s. He worked in research and development for Nokia and later Siemens. There, he led a team that architected the Pan-European ATM (asynchronous transfer mode) network, a Tier-1 ISP that served as Europe's backbone for global Internet access. He also helmed teams that successfully deployed GSM (global systems for mobile communications) for the Dutch railway system as well as launching Europe's first 3G UMTS (universal mobile Telecommunications system) network. Alam was born in Dhaka, Bangladesh, but moved to the Netherlands after receiving a scholarship to attend a prestigious boarding high school. Later, he was awarded degrees in electrical engineering from Holland's renowned Delft University of Technology, with a focus on personal mobile and radar communications.

Dr. K.-C. Chen received B.S. from the National Taiwan University in 1983, M.S. and Ph.D from the University of Maryland, College Park, United States, in 1987 and 1989, all in electrical engineering. From 1987 to 1998, Dr. Chen worked with SSE, COMSAT, IBM Thomas J. Watson Research Center, and National Tsing Hua University, in mobile communications. Since 1998, Dr. Chen has been with the Graduate Institute of Communication Engineering and Department of Electrical Engineering, National Taiwan University, Taipei, Taiwan, ROC. He was appointed as the Irving T. Ho Chair Professor from 2007 to 2008, and the Director of the Graduate Institute of Communication Engineering, and Director of Communication Research Center, 2009–2012. Dr. Chen is now Distinguished Professor and Associate Dean for academic affairs, College of Electrical Engineering and Computer Science, National Taiwan University. He was visiting Hewlett-Packard Laboratories in California, USA, during 1997 and a Guest Professor at the Delft University of Technology, Netherlands, 1998, Aalborg University, Denmark, 2008, and Visiting Scientist at the Research Laboratory of Electronics, Massachusetts Institute of Technology, 2012–2013 and 2014, SKKU Fellow Professor, Korea, 2013–2014. Dr. Chen was adjunctly appointed by the Executive Yuan Science and Technology Advisory Group to plan Taiwan's communication and networking technologies during 1998–2002, including telecommunication deregulation, cellular/fixed-network licensing, international trade negotiation, and facilitation of NCC under the authorization of Premiere. Dr. Chen is actively involved in the technical organization of numerous leading IEEE conferences, including the Technical Program Committee Chair of 1996 IEEE International Symposium on Personal Indoor Mobile Radio Communications, TPC co-chair for IEEE Globecom 2002, General co-chair for 2007 IEEE Mobile WiMAX Symposium in Orlando, USA, 2009 IEEE Mobile WiMAX Symposium in Napa Valley, USA, the IEEE 2010 Spring Vehicular Technology Conference, 2011 IEEE Online Conference on Green Communications, WPMC 2012, and many others. He has served editorship with the following prestigious international journals: IEEE Transaction on Communications, IEEE Communications Letters, IEEE Communication Surveys, IEEE Personal Communications Magazine, International Journal of Wireless Information Networks, IEEE Journal on Selected Area in Communications

(5 issues), IEEE Journal on Selected Topics in Signal Processing, IEEE Wireless Communications, ACM/Blatzer Journal on Wireless Networks, Wireless Personal Communications, Wireless Communications and Mobile Computing, Frontier of Communication and Information Theory, PHYCOM, etc. Dr. Chen founded IEEE Workshop on Social Networks and IEEE Workshop on Smart Grid Communications. He has been a voting member for IEEE 802.11 (wireless LANs), IEEE 802.15 (Wireless Personal Area Networks), IEEE 802.14 (HFC modem), IEEE 802.16 (WiMAX) international standard working groups, and participating US TIA45.5 CDMA Cellular standard, ETSI SMG2 cellular standard, and ITU-R TG8/1 IMT-2000 (3G) standard, ETSI 3GPP, and was Vice Chair WWRF SIG3 2006-7. He has authored and co-authored over 200 IEEE/ACM technical papers, 20 granted/pending US patents, a few book chapters, and 3 books Cognitive Radio Networks (with R. Prasad) by Wiley 2009, Mobile WiMAX (ed. with R. DeMarca) by Wiley 2008, and Principles of Communications by River 2009. Dr. Chen was elected as an IEEE Fellow in 2007 Class (special report by IEEE Spectrum), one of the Ten Outstanding Young Engineers in 1994, one of the Ten Outstanding Young Persons (the most prestigious achievement award for people under age 40 in Taiwan) in 1996, NSC Excellent Research Award in 2000, Outstanding Engineering Professor in 2002, etc. He was invited as a speaker in the United Nation ITU TELCOM 95 Technology Summit, Asia TELCOM 97 Strategy Summit, and keynotes in various international conferences in recent years. He also led APEC Telecommunication Working Group WTO Implementation task group with 19 member economies. Dr. Chen has served in IEEE, such as the IEEE Communication Society Asia Pacific Board Director during 2002–2003, IEEE VTS Fellow evaluation committee 2007–2012, IEEE Fellow committee 2013–2014, IEEE VTS Distinguished Lecturer 2012–2014, IEEE ComSoc Social Networks sub-committee chair since 2010, IEEE ComSoc Emerging Technology Committee Member 2013–2015, IEEE Signal Processing Society Big Data SIG member 2014–2016. His technology has been adopted in the IEEE 802.11 wireless LANs, Bluetooth 2.0 and beyond, IEEE 802.15, 3GPP LTE (i.e. 4G wireless communications) and LTE-A. Dr. Chen co-authored IEEE papers to receive 2001 ISI Classic Citation Award, IEEE ICC 2010 Best Paper Award, 2010 IEEE GLOBECOM GOLD Best Paper Award, and 2014 IEEE Jack Neubauer Memorial Award. Dr. Chen received 2011 IEEE ComSoc Wireless Communication Recognition Award. His research interests include data analytics, wireless communications and network science, particularly in 5G wireless and cyber-physical systems, large wireless networks, inference on big networked data, and biochemical molecular communications.

3

Experimental Activities to Support Future Space-based High Throughput Communication Infrastructures

Tommaso Rossi[1], Marina Ruggieri[1],
Giuseppe Codispoti[2] and Giorgia Parca[2]

[1]CTIF-Italy, University of Rome "Tor Vergata", Rome, Italy
[2]Italian Space Agency, Italy

Abstract

High Throughput Satellite systems still offer much less bandwidth per user with respect to terrestrial broadband networks. In order to reach a very high throughput towards "terabit connectivity", an important breakthrough is needed in terms of bandwidth availability which can be offered by the use of Extremely High Frequency bands (30–300 GHz). In particular, Q-V band (30–50 GHz) and W-band (70–110 GHz) offer very promising perspectives, being not used for commercial systems and offering a large part of the spectrum allocated for satellite services. However, EHF band transmission through the atmosphere is subject to absorption and dispersive effects in amplitude and phase that have to be carefully investigated. Moreover, some mitigation techniques have to be analyzed and properly tuned in order to realize an efficient transmission. In this section, the current experimental activities to support future EHF satellite communications are reported, with reference to the Italian Space Agency Q/V-band satellite communication experimental campaign.

Keywords: Satellite communications, terabit connectivity, extremely high frequencies, propagation impairments mitigation techniques, ACM, up-link power control, smart gateways.

Role of ICT for Multi-Disciplinary Applications in 2030, 59–76.

3.1 Introduction and Motivations

In the framework of modern Information Society, broadband satellite communication systems, with their global access and broadcasting capabilities, are assuming an increasing relevance. The future broadband-distributed satellite user access services will be developed with the main goal to support the so-called "terabit connectivity" (to support the increasing bitrate requirements), maintaining the user-capacity cost comparable to the one of terrestrial solutions. This goal can be reached through the exploitation of novel frequency bands that are currently not used for commercial services [1]. In particular, the frequency range that extends from 30 to 300 GHz, called Extremely High Frequencies (EHF), is an uncrowded part of the spectrum that can be effectively used to support very wide bandwidth channels in which information can be transmitted at a high rate [2]. Moreover, in addition to the large bandwidth availability, the use of millimeter waves leads to smaller antenna size for a fixed gain, or conversely, higher antenna gain for a fixed size with respect to lower frequencies waves; millimeter waves allow also RF equipment miniaturization.

The drawback of radiowaves at these frequencies are the strong impairments caused by the lower part of the atmosphere (troposphere), hence research activities on techniques for propagation impairment mitigation (PIMT) are needed, which are able to dynamically adapt the system to the channel conditions; in particular, Adaptive Coding and Modulation (ACM), Data Rate Adaptation (DRA), up-link power control, spatial diversity (both using classical site diversity approach or "smart gateways" approach), and on-board adaptive power allocation can be efficiently adopted to improve EHF satellite systems performance.

Currently, the use of Ka-band (30 GHz) is the benchmark for broadband satellite communications commercial application, while Q/V band (40–50 GHz) is under scientific investigation through European experimental campaigns; moreover, great scientific interest is pointed toward W-band (70–90 GHz) [3, 4]. In this framework, the new generation of High Throughput Satellites (HTS) for broadband distributed user access is strictly connected to the use of "beyond Ka-band" frequencies. As a matter of fact, HTS are bandwidth limited; hence, to reach a very high throughput, bandwidth efficient modulation and coding schemes have to be used and an important breakthrough is needed in terms of bandwidth availability [5].

Current HTS are systems exploiting multi-beam coverage to increase frequency reuse and dimensioned to support an aggregated total capacity of tens of Gbps over a coverage area as large as Europe, with a single spot capacity of hundreds of Mbps. In order to be an active part of the future 5G network, next generation of HTS systems shall be able to satisfy demanding user bit rate requirements. The first step to reach the important breakthrough needed in terms of bandwidth availability is the development of HTS system architectures based on the use of Ka-band for the user link, to maintain the user terminal compatibility with the current system, and Q/V-band (or beyond) for the feeder link. This solution is particularly attractive because the whole Ka-band spectrum (that is currently allocated both for feeder and for user links of the HTS systems) will become available for the user link segment (even if the use of some part of this spectrum requires coordination with other telecommunication systems). The use of both Q/V-band and Ka-band will also lead to the minimization of the number of terrestrial hubs, whose development and management is the main HTS systems cost.

In this frame, an important European experimental campaign is currently conducted by the Italian Space Agency (ASI) and the European Space Agency (ESA) with the objective to assess the use of Q/V-band frequencies for satellite communications. ASI has a proven expertise in the scientific research on very high-frequency satellite communications, having funded both operative systems for Ka and Q/V bands propagation analysis, as Italsat F1 [6], and advanced studies for the development of W-band (75–110 GHz) satellite communication systems [7, 8]. In 2004, ASI started to fund a program for the assessment of Q/V band satellite communications [9, 10].

In 2006, ESA invited the Member Agencies to propose innovative experiments that could be realized through technological payloads hosted as piggy-back aboard the new Alphasat big GEO satellite. ASI proposed an innovative Q/V-band payload able to support satellite communication and propagation experimental activities. Following several technology studies and preliminary accommodation activities, this payload has been selected as one of the four hosted payloads, namely Technology Demonstration Payloads (TDPs), for flying on Alphasat. The payload development has been supported by ASI as a contribution to the Alphasat project which is executed by ESA in the framework of the ARTES 8 Telecom program. Thales Alenia Space Italy and Space Engineering were the prime contractors for the development of the payload, named TDP#5 [11].

The Q/V-band scientific mission has been conceived by ASI with the support of Prof. Marina Ruggieri from the University of Roma Tor Vergata as Principal Investigator (PI) for the Communication Experiment, Prof. Carlo Riva and the late Prof. Aldo Paraboni from Politecnico di Milano as PI of the

Propagation Experiment. PIs provided ASI the scientific requirements for the development of Alphasat system and are responsible for the execution of the experiments. The ASI Q/V band payload was later renamed "Aldo Paraboni" Payload in memory of the late Prof. Paraboni.

The aim of the Communication Experiment is the test and optimization of PIMT over the Q/V band satellite channel, while the Propagation Experiment goal is the characterization of the propagation impairments in Q/V bands satellite channels. These two experimental campaigns are jointly conducted.

Alphasat was successfully launched on July 25, 2013, from the European Spaceport in Kourou (French Guiana) via the Ariane 5 ECA rocket. The orbit is slightly inclined GEO (due to some operative requirements for the main commercial payload of Alphasat managed by Inmarsat) at the location 25° East. The In Orbit Test (IOT) campaign has been concluded in December 2013 and the experimental campaigns started at the end of February 2014.

ASI assumed the commitment to realize and deploy the ground terminals in Italy. The prime contractor for the development of ground terminals is Space Engineering. A first communication and propagation ground station is located in Tito Scalo (near Potenza, South of Italy), while a second ground station is located in Spinod'Adda (near Milan, North of Italy). An additional communication and propagation terminal is installed in Graz, Austria; this station has been developed by Johanneum Research Institute (under an ESA contract) which has been invited by ASI to join the Q/V band experiments. Several other small propagation terminals are located around Europe.

The Q/V band TDP payload is composed of two self-standing payloads: a communication Experiment payload including a two-channel transparent transponder operating in Q/V band (with cross-strapping capabilities and having a useful bandwidth of 10 MHz) and a Propagation Experiment payload including two beacons operating in Q and Ka bands.

The central frequencies of the two communication channels are: 47.9 and 48.1 GHz for the uplink and 37.9 and 38.1 GHz for the downlink. The polarization is linear vertical. The payload is capable to manage three beams; up to two beams are simultaneously active. The fixed beam pointed toward the ground station located in Tito Scalo can be always active, while the second active beam can be the one pointed toward Spinod'Addaor the one pointed toward Graz.

The operative frequencies of the propagation beacons are 39.402 GHz and 19.701 GHz. The Q-band polarization is linear 45° tilted, while the Ka-band one is linear vertical. The two un-modulated signals are coherent (generated starting from a common oscillator).

The Communication Experiment payload transponder can be configured in cross-mode or in loop-back. In the first configuration, the received beam of one transponder serves the same geographical area of the other transponder transmit beam, and vice versa (each channel is received and transmitted on different antenna coverage). In the second configuration, each transponder is connected in transmission and reception to the same beam (each channel is received and transmitted on the same antenna coverage). With this configuration, the payload is capable to realize full-duplex communication between the station located in Tito Scalo and, alternatively, the station located in SpinoD'Adda or the one located in Graz.

The two Italian ground stations are equipped with a large Q/V band antenna (4.2 m), the high-power amplifier element is an Extended Interaction Klystron (EIK) with a nominal power of 50 W (and a peak of 200 W). The ground stations are also equipped with radiometers, pluviometers, and meteorological ancillary instruments (surface temperature, relative humidity, and atmospheric pressure).

The system architecture is reported in Figure 3.1. The experimental activities are managed through two control centers: one devoted to communication scientific experiments and another devoted to propagation ones. The first

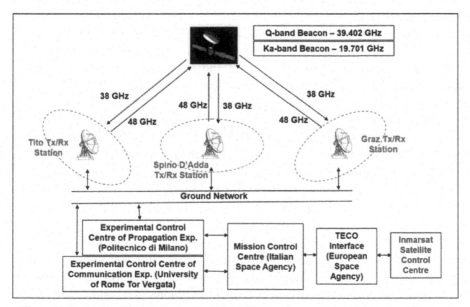

Figure 3.1 Alphasat Aldo Paraboni payload Q/V band experiments system architecture.

one is located at the University of Rome Tor Vergata, while the second is located at the Politecnico di Milano. Experimental plans generated in these centers are collected by a Mission Control Centre (MCC) that configures the ground terminals and delivers the satellite configuration requests to the Inmarsat Satellite Control Centre through an ESA interface called TDPs ESA Coordination Office (TECO).

The baseband communication section is based on the DVB-S2 (Digital Video Broadcasting second generation) standard [12, 13]. DVB-S2 is a standard for broadcasting services, interactive services, and data content distribution, developed by the Digital Video Broadcasting (DVB) Project in 2003.

3.2 Description of Q/V-Band Satellite Communication Experimental Activities and First Results

In this section, the activities of the ASI Q/V-band satellite communication experiments will be introduced together with some preliminary experimental results.

The PIMTs that will be object of the communication experiment are: Adaptive Coding and Modulation (ACM), up-link power control (UL-PC), and space diversity.

ACM experiments have the objective to design and optimize control algorithms for the specific 40–50 GHz communication channel. The goal is the near real-time optimal selection of the best ACM modulation and coding scheme (ModCod) for the particular channel conditions providing the best combination of spectral efficiency and Bit Error Rate/Frame Error Rate (BER/FER) [14]. ULPC experiment will optimize the adaptive change of the power transmitted by the ground terminal on the basis of link attenuation in order to maintain a constant power flux density at the satellite input. Space diversity experiments are devoted to the identification of optimal spatial re-routing of the radio path around the source of the fading [15].

These adaptive transmission techniques require a link quality estimator able to correctly identify propagation channel conditions. These conditions are evaluated with a closed-loop approach using different metrics as: Received Signal Strength (RSS), Carrier-to-Noise Ratio (CNR), Signal-to-Noise Ratio (SNR), Error vector magnitude (EVM), Bit Error Rate (BER), and Frame Error Rate (FER). On the other hand, an open-loop approach can be adopted to estimate link quality using the data gathered from the ancillary equipment, such as the radiometer data and the meteorological station data or coming from the Q-band beacon receiver. Moreover, measurements coming from closed-loop and open-loop control approaches can be jointly used to increase channel estimation performance. Channel estimation error and delay can result in low performance if the channel adaptive PIMT is not designed properly; the effects of channel estimation error and delay will be evaluated during the experiments.

Some of the preliminary results of channel estimation experiments are reported in Figure 3.2. Here a data-aided SNR estimator has been characterized in terms of standard deviation; in particular, an estimator based on pilot symbols inserted within the DVB-S2 frame has been used. The curves report the estimator standard deviation as a function of channel SNR, for different lengths of average windowing. The windowing is required to reduce estimation fluctuation (the estimation is performed on a frame basis). It can be noticed that there is a slight increase in the standard deviation for high SNR levels; this is due to the fact that the standard deviation is not only a function of the SNR value, but also a function of the total number of pilot symbols used for estimation; however, the number of useful pilot symbols in the frame decreases as the modulation spectral efficiency increases. Since high-order modulation schemes are used for high level of SNR, the number of pilot symbols used for the estimation decreases as the SNR grows, thus reducing the estimator accuracy. As expected, the windowing reduces the SNR estimator variance. A windowing length higher than 25 samples guarantees a SNR estimation

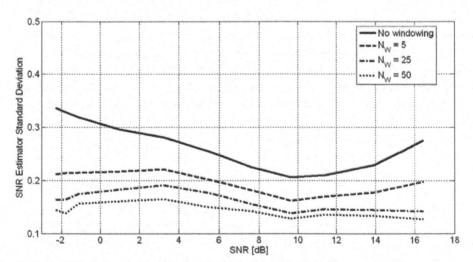

Figure 3.2 SNR estimation standard deviation for different average windowing lengths.

variance lower than 0.2 for the whole SNR operational range. It is also evident that a large windowing length reduces the effects of the decrease in the estimation accuracy for high-order modulation. Complete results can be found in [16].

As introduced previously, the transmission scheme is based on DVB-S2 standard; hence, a total of 28 modulation and coding schemes (ModCod) can be used for transmission on the basis of channel conditions. This ACM technique provides a lot of flexibility but, on the other hand, increases the physical layer selection algorithm complexity. The correct selection of a subset of useful ModCod is of capital importance to optimize the physical layer adaptation technique; in this frame, in the first part of the experimental campaign, the FER curves, as a function of SNR, have been identified for all the ModCod provided for the DVB-S2 standard. Figure 3.3 shows the experimental results obtained for analyzing real communication data from the "Aldo Paraboni" payload. In particular, FER vs SNR curves are reported for the useful subset of DBV-S2 ModCod.

The basic concept of ACM is the dynamic adaptation of the coding and modulation scheme with the channel conditions, privileging more spectrally efficient ModCod in clear sky and more robust schemes if poor channel conditions prevail. In particular, the implementation of such technique requires the estimation of the average SNR (inside a specific data window) for the comparison with a thresholds vector, whose elements correspond to a specific ModCod. Hence, in the following part of the experimental activities, the operative SNR thresholds used to operate switching between ModCod have been identified in order to satisfy the FER service requirements and optimize the use of system resources. These thresholds are selected to optimize two conflicting requirements: avoid frame losses, and increase spectral efficiency. The former requirement lead to the use of more protected ModCod (low modulation order and strong channel coding), and the latter requirement lead to the use of ModCod with high spectral efficiency (high modulation order and low redundancy channel coding).

Both ACM control logic based on fixed thresholds and hysteresis control loop have been optimized during this first part of communication experiments. The first logic is based on the use of a static fixed threshold for every ModCod; the second logic is based on the use of two thresholds in a hysteresis loop in order to minimize frame losses and reduces the amount of reverse link signaling. This hysteresis loop logic can be included in the ACM mode switching algorithm to prevent undesired oscillations of the transmitting ACM mode in case of SNR jitter between adjacent SNR thresholds.

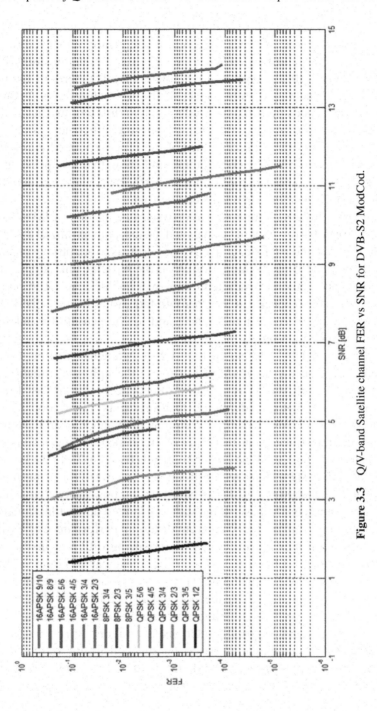

Figure 3.3   Q/V-band Satellite channel FER vs SNR for DVB-S2 ModCod.

Accounting for the long round trip delay typical of a GEO satellite link, this may lead to undesirable mode switching, and consequently either reduction in data rates or increase in BER. The first test on real communication data has been performed with the purpose to evaluate the advantages of the use of hysteresis ACM technique with respect to the fixed thresholds method. The comparison has been performed introducing two different metrics, communication success probability, and average spectral efficiency, which in turn allowed to identify the optimal values of the proposed ACM algorithm parameters. The results are very encouraging and show that it is possible to optimize the hysteresis control loop so that its performance is better than the static control loop. The complete results can be found in [14].

In the following phase of communication experiments, other PIMTs such as UL-PC and spatial diversity will be analyzed, tested, and optimized through Aldo Paraboni Q/V band payload. Objective of ULPC is to adaptively change the power transmitted by a ground terminal on the basis of link attenuation in order to maintain a constant satellite input power flux density. ULPC goals are: to improve system capacity, to compensate fade, to efficiently distribute power consumption, to reduce interference, and to compensate for antenna gain roll-off and mis-pointing.

Both open-loop ULPC and closed-loop ULPC strategies will be tested. The first one relies on the possibility to estimate the uplink attenuation at the transmitting ground station using some independent means, as downlink beacon measurement or radiometric measurements. Open-loop ULPC experiment will use the downlink beacon signals in Q and Ka bands; uplink attenuation will be estimated on the basis of frequency scaling technique. The effectiveness of the open-loop power control is based on both the correct detection of downlink signal fade and the frequency scaling of this value to the uplink frequency. Closed-loop ULPC is based on the possibility to evaluate the up-link attenuation on the basis of the self-transmitted down-link signal or the feedback coming from the received ground station. Performance of ULPC will be evaluated during experimental activities using different metrics as power control accuracy and maximum level of fade that can be compensated.

Space diversity PIMTs, including site diversity and smart gateway diversity techniques, are based on the concept of spatial re-routing radio links avoiding the source of deep fading; this can be done because rain is a phenomenon spatially and temporally intermittent and inhomogeneous.

The site diversity technique is based on the concurrent reception of the same signal by (usually) two ground stations few kilometers apart and connected via a terrestrial link; if the signal is heavily attenuated in one area, the signal received by the station with the best propagation conditions is used. The performance gain achieved using site diversity is dependent on the space and time correlation of rain events. The objective of site diversity is to maximize the service availability.

The smart gateway diversity technique which employs a number of gateway stations interconnected via terrestrial links is a very promising approach for broadband HTS systems. The objective is to realize a feeder link diversity scheme. The system architecture is based on the use of some GWs connected with a terrestrial fiber network so that it is possible to route feeder link data in a diversity manner to counteract deep fades on one (or more) gateway(s) [15]. This kind of technique is considered mandatory to reach the feeder link connection availability time required by the future HTS systems, but, on the other hand, the hubs must manage very complex handover procedures.

Space diversity techniques will be investigated during Q/V-band communication experiments. The experiment's main objective is to collect a large set of data to identify efficient station switching control procedure and diversity management logic, and analyze stations synchronization issues. The expected results will mainly concern to spatial diversity management.

3.3 Expected Impacts and Conclusions

The efficient use of EHF spectrum for satellite communications will lead to an important change in satellite-based services; as a matter of fact, the large bandwidth available at these frequencies will pave the way toward the terabit connectivity. This is of paramount importance in the future vision of an integrated terrestrial-satellite communication network where the resources are allocated by a network controller on the basis of new "softwarization paradigms" as Software Defined Networking (SDN) and Network Functions Virtualization (NFV). The challenge will be for satellite operators and ground system operators to work together to leverage these new capabilities into a range of new and cost-effective services. The SDN/NFV paradigms allows to develop networks and network functions in a cost-effective, flexible, and fast way. Programmability and virtualization are the key principles of SDN/NFV concept. Satellite communications, and in particular HTS access networks, can benefit from the adoption of SDN/NFV paradigm by many means: enabling fine-grained, fast-configuring services, lowering CAPEX and OPEX costs, enhancing the resiliency and the availability of the infrastructure, enabling fast-deployment of networks and network functions, and fast endpoint configurations [17]. The SDN/NFV progressive introduction into satellite and terrestrial networks also supports the developments of payloads with on-board processing architectures and advanced switching/routing capabilities.

The future generation of HTS, based on the use of EHF, will allow the globalization of gigabit/terabit data communication capabilities, dramatically improving the user capacity performance of current satellite-based systems. As a matter of fact, Internet access is a key driver for HTS systems, but there are other applications which will drive usage of HTS systems such as cellular backhaul for 3G, 4G, and future 5G mobile technologies when distances to cellular base stations make it cost-prohibitive using terrestrial means. Another interesting service provided by HTS is enterprise high-availability networking; a strong value driver for satellite networks is backup of terrestrial services, which ensures the highest availability even when disaster strikes. In addition, the satellite path can be used to instantaneously deliver bandwidth where and when it is needed.

The efficient use of PIMTs at EHF band requires an optimization process and a long test campaign, which is one of the main objectives of the ASI Q/V-band experimental campaign. The Alphasat "Aldo Paraboni" experiment allows us to perform, for the first time, communication experiments over a Q/V band satellite link with adaptive PIMT. In particular, ACM techniques based on DVB-S2 standard, together with up-link power control and space diversity techniques, are currently under testing and optimization. The results and the experience acquired from these experiments will allow a more effective use of the Q/V spectrum and will drive the future development of HTS systems.

References

[1] Cianca, E., Rossi, T., Yahalom, A., Pinhasi, Y., Farserotu, J., and Sacchi, C. (2011). EHF for Satellite Communications: The New Broadband Frontier. *Proc. IEEE* 99(11).

[2] Electronic Communication Committee. "The european table of frequency allocations and applications in the frequency range 8.4 kHz to 3000 GHz", ERC Report.

[3] Sacchi, C., Rossi, T., Ruggieri, M., and Granielli, F. (2011). Efficient Waveform Design for High-Bit-Rate W-band Satellite Transmissions. *IEEE Trans. Aerospace Electronic Syst.* 47(2), Apr., ISSN: 0018-9251.

[4] Stallo, C., Cianca, E., Mukherjee, S., Rossi, T., De Sanctis, M., and Ruggieri, M. (2013). UWB for multi-gigabit/s communications beyond 60 GHz. *Telecommun. Syst. J.* 52, ISSN: 1018-4864.

[5] De Sanctis, M., Cianca, E., Rossi, T., Sacchi, C., Mucchi, L., and Prasad, R. (2015). Waveform design solutions for EHF broadband satellite communications. *IEEE Commun. Magazine* 53, 18–23.

[6] Polonio, R., Riva, C. (1998). ITALSAT propagation experiment at 18.7, 39.6 and 49.5 GHz at SpinoD'Adda: three years of CPA statistics. *IEEE Trans. Antennas Propagat.* 46.

[7] Jebril, A., Lucente, M., Ruggieri, M., Rossi, T., Iera, A., Molinaro, A., Pulitano, S., Sacchi, C., Zuliani, L. (2008). Experimental missions in W-band: a small LEO satellite approach. *IEEE Syst. J.* 2 90–103.

[8] Jebril, A., Lucente, M., Re, E., Rossi, T., Ruggieri, M., Sacchi, C., Dainelli, V. (2007). "Perspectives of W-band for space communications", in *IEEE Aerospace Conference 2007*, Big Sky, Montana.

[9] Rossi, T., Cianca, E., Lucente, M., De Sanctis, M., Stallo, C., Ruggieri, M., Paraboni, A., Vernucci, A., Zuliani, L., Bruca, L., Codispoti, G. (2009). "ExperimentalItalian Q/V band satellite network", *IEEE Aerospace Conference*, Big Sky, Montana, March, ISBN: 978-1-4244-2621-8.

[10] Cianca, E., Stallo, C., Rossi, T., De Sanctis, M., Vernucci, A., Cornacchini, C., Bruca, L., Lucente, M. (2008). "Transponders: effectiveness of propagation and impairments mitigation techniques at Q/V band", in *Globecom 2008–IEEE Workshop/EHF-AEROCOMM, "Exploitation of Higher Frequency Bands in Broadband Aerospace Communications"*, November 30–December 4, New Orleans, LA, USA, ISBN: 978-1-4244-3061-1.

[11] Ruggieri, M., Riva, C., De Sanctis, M., Rossi, T. (2012). "Alphasat TDP#5 mission: towards future EHF satellite communications", in *IEEE 1st AESS European Conference on Satellite Telecommunications (ESTEL)*, pp. 1–6, ISBN: 978-1-4673-4687-0.

[12] ETSI TR 102 376, "Digital Video Broadcasting (DVB)—User guidelines for the second generation system for Broadcasting, Interactive Services, News Gathering and other broadband satellite applications (DVB-S2)", V1.1.1, 2005.

[13] ETSI EN 302 307, "Digital Video Broadcasting (DVB)—Second generation framing structure, channel coding and modulation systems for Broadcasting, Interactive Services, News Gathering and other broadband satellite applications (DVB-S2)", V1.2.1, 2009.

[14] De Sanctis, M., Rossi, T., Rizzo, L., Ruggieri, M., and Codispoti, G. (2015). "Optimization of ACM algorithms over Q/V-band satellite

channels with the Alphasat Aldo Paraboni P/L", in *IEEE Aerospace Conference 2015*, Big Sky, Montana, 7–14 March.

[15] Rossi, T., Maggio, F., De Sanctis, M., Ruggieri, M., Falzini, S., and Tosti, M. (2014). "System analysis of smart gateways techniques applied to Q/V high throughput satellites throughput satellites", in *IEEE Aerospace Conference, 2014*, Big Sky, Montana, March 1–8.

[16] Rossi, T., De Sanctis, M., Maggio, F., Ruggieri, M., Codispoti, G., and Parca, G. (2015). "Q/V-band satellite communication experiments on channel estimation with alphasat aldo paraboni P/L", in *IEEE Aerospace Conference 2015*, Big Sky, Montana, 7–14 March.

[17] Rossi, T., De Sanctis, M., Cianca, E., Fragale, C., Ruggieri, M., and Fenech, H. (2015). "Future space-based communications infrastructures based on high throughput satellites and software defined networking", in *2015 IEEE International Symposium on Systems Engineering (ISSE)*, pp. 332–337. doi: 10.1109/SysEng.2015.7302778.

Biographies

T. Rossi received his University Degree in Telecommunications in 2002, MSc Degree in "Advanced Communications and Navigation Satellite Systems" in 2004, and PhD in Telecommunications and Microelectronics in 2008 at the University of Rome "Tor Vergata" where he is currently an Assistant Professor (teaching Digital Signal Processing and Image and Video Compression). His research activity is focused on Space Systems, EHF Satellite and Terrestrial Telecommunications, Satellite and Inertial Navigation Systems, Digital Signal Processing for Radar, and TLC applications. He is currently Co-Investigator of the Italian Space Agency Q/V-band satellite communication experimental campaign realized through the Alphasat "Aldo Paraboni" payload.

M. Ruggieri is Full Professor of Telecommunications Engineering at the University of Roma "Tor Vergata" and therein member of the Board of Directors. She is the co-founder and Chair of the Steering Board of the interdisciplinary Center for Teleinfrastructures (CTIF) at the University of Roma "Tor Vergata". The Center, that belongs to the CTIF global network, with nodes in the USA, Europe, and Asia, focuses on the use of the Information and Communications Technology (ICT) for vertical applications (health, energy, cultural heritage, economics, and law) by integrating terrestrial, air and space communications, computing, positioning, and sensing. She is Principal Investigator of the 40/50 GHz TPD#5 Communications Experiment on board, the European Alphasat satellite (launched on July 2013).

She is Sr. past President of the IEEE Aerospace and Electronic Systems Society (2010–2011), where she served as a Member of Board of Governors since 2000 for three terms; Founder and Chair of the Space Systems Panel (2002–2010); Editor for Space Systems in the Transactions on Aerospace and Electronic Systems (2001–present), Associate/Sector Editor and Assistant Editor of the Systems Magazine (2005–2009); International Director of Italy & Western Europe (2005–2014); and Chair of the N&A Committee (2012–2013). She is IEEE Division IX Director (2014–2015) and hence a sitting member of the Board of Directors and the Technical Activities Board. She has been a member of the TAB Strategic Planning Committee (2011–2014), and TAB Representative in the Women in Engineering Committee (2011). She is a member of the IEEE Public Visibility Committee (2015), Governance Committee (2015), Fellow Committee (2015), and TAB Representative in the IEEE Awards Board (2014–2015).

She is Vice President of the Roma Chapter of AFCEA; proboviro of the Italian Industries Federation for Aerospace, Defense and Security (AIAD); and member of the Technical-Scientific Committee of the Center for Aeronautical Military Studies.

She received: 1990 Piero Fanti International Prize; 2009 Pisa Donna Award as women in engineering; 2013 Excellent Women in Roma Award; and Excellent and Best Paper Awards at international conferences.

She is an IEEE Fellow. She is author/co-author of 320 papers, 1 patent, and 12 books.

G. Codispoti received a degree in Electrical Engineering from the University of Calabria, Italy, and a Master of Science degree in Electrical Engineering from the California Institute of Technology in Pasadena, USA. During his graduate studies, he was involved in class projects at the Caltech/ NASA Jet Propulsion Laboratory in Pasadena with the responsibility of the communication aspects. From 1993 to 2000, he was with AleniaSpazio, Rome (now Thales Alenia Space Italia) at the "On board Active Antennas Department" as designer, project and program manager in either Telecommunication and Remote Sensing programs. In March 2000, he joined ASI, the Italian Space Agency, where he has been involved in projects regarding microgravity, remote sensing and telecommunications. At the moment, he works in the Telecommunication and Integrated Application Division and he is the responsible for the Q/V Band Program of the Agency.

He has been appointed as delegate and expert of the Italian Government in delegations of international bodies such as European Space Agency (ESA) and United Nations Organization (UNO). He is a member of Technical and Scientific Committees of public Foundations. He is a tutor of Ph.D students of Italian universities.

G. Parca, she holds Master's Degree in Telecommunications Engineering (2006) and PhD in Telecommunications and Microelectronics Engineering (2010) at the University of Rome Tor Vergata, Electronic Engineering Department. Main research topics have been optical wired, wireless, inter-satellite high-speed networks. After a Post-Doctoral fellowship at the Portuguese Telecommunications Institute, on optical telecom systems and devices for data/image processing, she is Research fellow at the Italian Space Agency since 2013, with the Telecommunications and Navigation division. Main research areas are on Ka and Q/V band satellite communications and maritime surveillance applications. She is author of several papers on international journals and conferences proceedings.

4

High-Capacity Ground-, Air, and Space-based ICT Networks for Communications, Navigation, and Sensing Services in 2030

Enrico Del Re[1], Simone Morosi[1], Luca Simone Ronga[2]
and Sara Jayousi[1]

[1]Department of Information Engineering, University of Florence,
Firenze, Italy
[2]CNIT, Research Unit of Florence, Firenze, Italy

Abstract

An integrated terrestrial-satellite system is proposed to provide the capabilities of communication, localization, and sensing for the innovative multidisciplinary services of the future pervasive Internet *Ecosystem*. They include services to enhance the quality of our everyday life as well as applications in critical situations and environments. Such a scenario requires the synergistic use of communication, localization, and sensing/monitoring techniques provided by a meshed heterogeneous architecture based on terrestrial and satellite segments to afford remarkable quality and performance to the future society based on the Internet of Things (IoT).

Keywords: Integrated terrestrial-satellite system, communications, localization, monitoring, sensing, cognitive networks, context awareness, internet of things.

4.1 Introduction

Recent technological and societal changes move our society toward a future where telecommunication platforms will be integrated with heterogeneous

Role of ICT for Multi-Disciplinary Applications in 2030, 77–94.

Societal Challenges
Health, demographic change & wellbeing (ICT)
Food security, sustainable agriculture,
& the bio-based economy (ICT)
Secure, clean & efficient energy (ICT)
Smart, green & integrated transport (ICT)
Climate action, resource efficiency,
& raw materials (ICT)
Inclusive, innovative & reflective societies (ICT)
Secure societies (ICT)

Figure 4.1   Guide on ICT-related activities in H2020 WP 2016-17.

Source: https://www.ideal-ist.eu/sites/default/files/Guide_to_ICT-related_activities_in_H2020WP2016-17_0.pdf

systems of localization and sensing capabilities, and intelligent objects will cooperate as in human social networks with the goal of providing multi-disciplinary services to the people and organizations at different levels of complexity, as foreseen in the paradigm of the Internet of Things (IoT) or even of the more challenging Internet of Everything (IoE). To achieve such a visionary scenario, each component has to be designed (and often redesigned) taking into account a remarkable level of generality and universality that characterizes the new system entities. New architectures and infrastructures have to be defined considering a plurality of services and applications with a low (ideally with no) reset effort. Such platform can support applications and services according to the objectives defined by the EU Horizon 2020 research and innovation program [1]. An interesting view of the possible ICT-related applications in societal challenges of everyday life of European citizens is shown in Figure 4.1.

The success of this strategy depends on a plurality of key factors that are briefly described in the following:

- The approach will have to be adopted for a very large set of possible future services, even if their requirements are rather diversified and even divergent;
- The proposed system will be characterized by distributed intelligence (sensing, learning, decision, and action), reconfigurability, adaptability, context awareness (physical, environmental, situational, and geographic), energy efficiency, and security; moreover, it will be able to exploit the benefits of clouds of users;

- The communications, both within the system and between the system and the external world, will have to be highly efficient and to deliver all the necessary information in due time, where needed, with required quality of service (QoS) and energy efficiency;

The reconfigurable heterogeneous infrastructure (the platform) that will be able to effectively cope with all these key issues will have to integrate localization and sensing/monitoring capabilities with terrestrial and satellite communication subsystems and intelligent objects and networks. As a result, the quality of life of the citizens and users will be improved in all scenarios of future IoE, such as for example but not limited to:

- Health, demographic change and wellbeing,
- Food security, sustainable agriculture, and bio-based economy,
- Secure, clean, and efficient energy,
- Smart, green, and integrated transport,
- Climate action, environment, resource efficiency, and raw materials,
- Europe in a changing world – Innovative, inclusive, and reflective societies,
- Secure societies – Protecting freedom and security of Europe and its citizens.

Moreover, the platform will be highly reconfigurable and adaptable to different application contexts: as an example, the system should be able to update the configuration parameter and the radio interface of a handheld device (e.g. a Smartphone) to provide the right information when and where needed.

This visionary objective demands the availability of a ubiquitous and flexible 5G infrastructure and of a complementary and seamlessly integrated terrestrial and satellite networks with very high capacity and great flexibility to meet the challenging requirements of the future Internet-of-Everything (IoE). In order to allow the implementation of such a system, the provision of a heterogeneous satellite-terrestrial architecture based on the paradigm of Software-Defined-Network/Network-Function-Virtualization (SDN/NFV) and the use of new Q/V and W bands for the satellite connections are an appropriate approach. The full reconfigurability of the heterogeneous network asks for frequency-agile transmission systems for capacity increase and defines the higher-layer network according to the new paradigms of access virtualization. This approach is in agreement with the recent trend of separation of the control plane from data transport in SDNs and with the progressive "softwarization" of all network layers, including the physical one. The migration toward NFV allows fine-grained definition of available services on a per-user basis, introducing new cross-tier interactions among information domains (user position, user context, user environment, etc.). Enabling NFV into satellite segment provides operators with suitable tools and interfaces in order to establish end-to-end fully operable virtualized satellite networks for third-party operators/service providers. SDN control center management is the flexible solution for non-proprietary adaptation to changing requirements and hardware products with ease upgradeability (in terms of both time and costs). The 5G terabit connectivity, achieved by integrating in seamless and cognitive manner 5G terrestrial and satellite network segments operating in the millimeter wave bandwidths, optimizes system resources and reduces the number of terrestrial hubs (whose development and management is the main system cost). The application of SDN/NFV paradigm to the HUB layer provides great benefit and simplifies the complex tasks to implement the Propagation Impairments Mitigation Techniques (PIMT) for the efficient use of EHF satellite links. "Smart gateways" (SGWs) are one of the most challenging PIMT, a spatial diversity scheme for the feeder link using a pool of synchronized GWs connected with terrestrial fiber network to route feeder link data traffic to counteract deep fades on one or more GWs. Virtualization technologies applied to the HUB layer will pave the way to an efficient implementation of these concepts for a "Virtual Smart Hub" where all resources will be virtualized.

We will propose a possible overall flexible system architecture for a high capacity ground-, air-, and space-based ICT networks for communications, navigation, and sensing services for future IoE.

This chapter is organized as follows. Section 4.2 is devoted to the description of the potential application scenarios of future integrated services and to the identification of the high-level requirements. Section 4.3 describes the flexible heterogeneous baseline architecture, highlighting the enabling technologies to support the provision of services for enhancing the quality of life. Section 4.4 presents some of the new paradigms and challenges for the development of advanced, highly reconfigurable, reliable, and secure systems. Finally, Section 4.5 concludes the chapter.

4.2 Toward Future Integrated Services

The objective of this section is to introduce future integrated services in some of the main ICT application sectors: starting from an overview of potential scenarios, the general requirements for the definition of a baseline system architecture are identified.

4.2.1 Potential Scenarios

A brief description of future potential integrated services is provided, focusing on the following application contexts:

- E-Health and Well-being,
- Public Safety,
- Mobility for Smart Cities,
- Entertainment.

4.2.1.1 e-Health and well-being

The main trend in the e-health sector is to empower patients in the management of their health, reducing the cost of medical assistance as well as coping with the increase in chronic diseases and the aging population. Smart devices with increased processing power, wearable technology, biosensors, and implantable devices combined with health application tools will enable the monitoring of a patient health status, remote diagnosis, and the self-control of chronic diseases. Anytime and everywhere (at home, in the hospital, in a rehabilitation structure, during a travel, outdoor, etc.), patient vital parameters can be monitored by smart objects interconnected with a medical assistance center. The user-transparent interactions among smart objects and the information exchange among the different entities (patient, doctors, family members) integrated with context-awareness (localization data) allow the provisioning of services ranging from well-being to chronic disease and emergency management.

Patients and caregivers (including specialists) are all part of a universal healthcare system: the former will be more responsible for their own health thanks to the active involvement in the management of their treatment and care, while the latter will act as a coordinator of treatment for patients. The processing of the collected data will automatically manage patients and request for specialist intervention as needed. Moreover, all the collected data will be used for the clinical studies, helping decision making and diagnosis.

4.2.1.2 Public safety

The development of new integrated system, including environmental monitoring, data transmission over satellite and terrestrial links, localization services based on advanced solutions for user-tracking, and innovative mechanisms for secure data exchange over wireless links, is leading rescue or safety team to adopt these technologies for guaranteeing public safety and security. In such a context, each user represents the destination of the alerting messages and the source of data coming from the interactions between the wearable/handled smart devices and the overall network and which transparently increase the contextual awareness of a particular event. In the next future, in order to help first responders to operate in critical contexts and to be guided by remote assistants (coordinator and control center), the transmission of the perceived reality (e.g. smell of burning substances, floor vibrations, etc.) will be one of the pursued innovative features, oriented to the achievement of an augmented reality.

4.2.1.3 Mobility for smart city

Environmental monitoring through roadside objects, traffic information based on a selected path in the navigator, establishment of relationships among the cars for useful information sharing (traffic, parking, accidents, etc.), and the trend of automated transport (self-driving cars) are some of the elements that will be part of the future system for mobility. Automotive manufacturers aim at the concept of a "connected vehicle" whose functions rely on underlying smart systems and networks for the information exchange, data analysis, and context aware solutions identification.

4.2.1.4 Entertainment and culture

Focusing on culture and entertainment contexts, and in particular, on the augmented reality technology in museums, the main trend is to interactively guide the visitor to a personalized cultural heritage experience through socio-personal interactions methodologies supported by multiple technologies. In the future, the possibility to recreate a full sensorial experience for a more effective perception of the artworks is foreseen thanks to the transmission of perceptions of all the five senses of a subject.

4.2.2 Requirements for Integrated Communications, Navigation, and Sensing Services

As highlighted in the description of the aforementioned scenarios, the seamless integration of the communication, navigation, and sensing features is one of the main objectives to be pursued for the development of advanced services.

The aim of this section is to identify a set of high-level requirements to be considered for the provisioning of such services. In details, due to the large variety of application contexts and the specific requirements for each service, the focus is on the definition of the general requirements for the design of a baseline architecture able to support a multiple services platform.

Aiming at translating the main users' needs to access the requested services into system requirements, the following features shall be considered:

- *Integration of heterogeneous technologies.* Existing technologies and systems shall be part of future architectures. User transparent integration of legacy and advanced systems is required for the exploitation of the different and complementary capabilities of each system and therefore for the development of personalized services based on user's specific needs.
- *Flexibility and scalability.* Future architectures shall be able to efficiently respond to unexpected events, which may affect the provisioning of services. Flexibility allows the system to self-adapt to the temporary context conditions in terms of network resources availability and environmental changes of specific location. Scalability can be seen as the system flexibility of the size of the users to be served.
- *Availability and reliability.* The users' increasing dependence on the ICT services asks for an always available network regardless of the user location. This allows users to access the desired data anytime and anywhere. Moreover, the continuous exchange of information and the big amount of data to be stored and processed in the cloud requires high reliable networks.
- *Reconfigurability.* The trend of providing highly customized services and the heterogeneity of the availability of network resources for the delivery of services (essentially based on the user location) requires the design of a high reconfigurable system. Reconfigurability shall be considered in terms of network resources allocation, network nodes functions, network access technologies usage, etc.
- *Security and Resilience.* Satisfying the user's mobility need means operating in an open access environment. Since multiple threats characterize the open wireless channel, the future architecture shall provide high security mechanisms for data transmission, distributed storage, and distributed access to personal data.

Future architectures shall be able to meet all the previous high-level requirements through the adoption of different technologies based on the specific service needs (e.g. 5G systems for high capacity networks, satellite communication for global coverage, higher availability, delay tolerant services, etc.). Moreover, the introduction of new paradigms (Software Defined Networking, Network Functions Virtualization, Cognition, Social Networking, Human

Bond Communications) in the design of the advanced systems will allow the development of innovative integrated services (Section 4.4).

4.3 Baseline System Architecture and Enabling Technologies

4.3.1 Baseline Architecture

As a matter of facts, the provision of innovative ICT services in the scenarios which have been described previously requires the integration of telecommunication entities and heterogeneous system architectures with localization and sensing capabilities and in which people and intelligent objects will cooperate as in social networks [2]. As a result, a pervasive communication *Ecosystem* is likely to be realized in the near future. In order to understand the features of the future heterogeneous networks, it is worth discussing the archetypical architecture, which is thoroughly described in Del Re et al. [2] (Figure 4.2).

In order to support the development of new advanced services also exploiting the multiple combination of the existing ones, the proposed system architecture is based on the integration of the aforementioned functionalities: localization, communication, and sensing/monitoring [3]. This flexible, scalable, and reliable heterogeneous architecture includes the satellite component as a key element. Satellite GNSS, communication, and monitoring systems are all parts of the proposed architecture together with different terrestrial components. This meshed architecture enables the cooperation among satellite systems and terrestrial ones with the goal to benefit from the capabilities of each system. In particular, the capabilities of each single subsystem can be improved by the exploitation of the features of the other ones, also in critical contexts. Moreover, a full interoperable architecture will be enabled by the distributed system intelligence and by suitable interfaces.

Even if the integration of services is currently possible only at client level (i.e. smart-phones, car computers, navigation devices) with a heavy limitation of the potentials of the services, a deeper level of integration will be afforded in the near future, also of the satellite component, so allowing to move services over networks, rather than moving clients over services and with a deeper level of integration. This objective could be afforded by the resorting to the separation of the control plane from data transport as in Software Defined Networks (SDN).

Nonetheless, since the implementation of applications is currently bound to the available access technology for each specific context, in the following

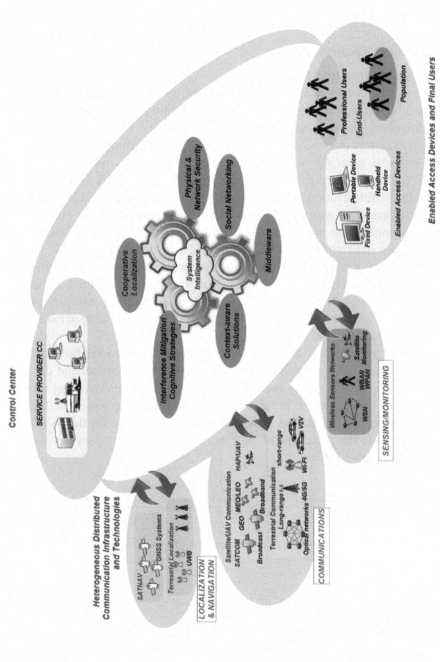

Figure 4.2 Baseline architecture proposed in Del Re et al. [2].

the most advanced enabling technologies which will permit the next leap forward are reviewed.

4.3.2 Enabling Technologies

4.3.2.1 5G systems and SatCom

Next generation wireless system, which is defined as fifth generation or 5G, will be the ubiquitous ultra-broadband network enabling the Future Internet (FI): as described in [4], it will enable "full immersive (3D) experience" enriched by "context information" and "anything or everything as a service (XaaS)" and will open the way to a world where the network has been reduced to a ubiquitous "pipe of bits". After the adoption of 5G systems, a revolution in the Information and Communications Technologies (ICT) field can be envisaged. It is worth underlining that, with respect to the current wireless communications, the future 5G communications shall guarantee the key performance indicator gains that are reported in Table 4.1 [4].

In this context, the issues of connection, security, mobility, and routing management will be effectively afforded. Full compatibility with current and future incremental 4G releases will be guaranteed by the possibility of instantiating any type of virtual architecture and installing any kind of network and service application efficiently.

In order to play an active role for the future 5G networks, the next generation of High-Throughput Satellite (HTS) systems for broadband distributed user access shall satisfy the user bit rate requirements reaching the "terabit connectivity" from the present tens of Gbps. Regenerative satellite with flexibility in the uplink and downlink connections, e.g. as described in [5], will provide the paradigm of 'Internet in the sky' necessary for the provision of the global IoE.

Table 4.1 5G key performance gains

Key Performance Indicator	Gain to Be Afforded
Throughput	1000× more in aggregate, 10× more at link level
Latency	1 ms for robot remote control or tactile Internet applications, below 5 ms for the download of 2–8K videos
Reliability	Ultra-high
Coverage	Suitable for a seamless experience
Battery lifetime	10× longer
Spectrum utilization	All spectra, from cellular bands to visible light

To manage efficiently the huge demand of user capacity and the propagation impairments, mitigation techniques required to realize an efficient EHF satellite transmission, a very attractive technique is the smart gateway and the adoption of the paradigm of SDN/NFV [6]. Smart GW for the feeder link is based on the use of a number of GWs connected by a terrestrial fiber network to route feeder link data by diversity availability to counteract occurrence of unavailable gateway(s). This operation requires the management of complex handover procedures and a precise localization of the mobile users.

4.3.2.2 Cooperative/assisted localization and UWB

Accurate localization of the user is nowadays very important, both for delivering new smart services (location-aware) as well as for outdoor and indoor navigation. An important support to enhanced-value location-based services can be identified in the integration of GNSS and other positioning techniques. In particular, GNSS involves several systems such as GPS, GLONASS, and the upcoming GALILEO, plus other regional navigation systems, so increasing the service availability in outdoor environment [7, 8].

Moreover, Peer-to-Peer Cooperative Positioning (P2P-CP) technique is able to improve the GNSS receiver performance in terms of availability, accuracy, and Time-To-First-Fix (TTFF) and can be an effective solution [9]: in P2P-CP context a terrestrial wireless communication link is required between cooperating users to allow the aiding information exchange; to this aim, short-range communication systems such as WiFi and Bluetooth can be used.

On the other hand, Dead Reckoning allows to estimate the user motion either in indoor or outdoor scenarios without any external infrastructure assistance, relaying only on a single terminal and its internal sensors [10]: nonetheless, this technique is often integrated with GNSS to improve the position accuracy when satellite signals are not available for a short time. In presence of hostile conditions, i.e., when the LOS to the satellites is partially or totally obstructed for a long time, the received signal strength might be too weak for an appropriate processing, leading the GNSS-based localization to degrade or even fail: in this context, stand-alone techniques may not afford the location services requirements.

Ultra-wide band (UWB) technology could be a good candidate for an accurate localization of smart devices. UWB can be also used successfully to enhance the accuracy of other positioning technologies. Diversity navigation employing multiple sensing technologies can overcome the limitation of individual technologies, in particular in harsh environments [11].

4.3.2.3 Wireless sensors networks

The realization of a monitoring system can be achieved through the implementation of a dedicated Wireless Sensor Network (WSN).

The WSN is composed of autonomous modules (with sensors for the detection of environmental parameters), which are characterized by low cost, small size, very low (down to no) maintenance effort, reduced environmental impact, and equipped with on-board intelligence.

This solution allows the monitoring system to acquire and process a signal and send the information to an operation and alert center via SMS: hence, it can be responsible for the safety and monitoring of areas subject to environmental risks.

As a matter of fact, a wireless sensor network for Environmental Monitoring Applications defines a prototype of Smart Grid: the goals of these networks can be extremely differentiated and, sometimes, contradictory, e.g., the responsiveness and the energy efficiency of the networks are opposite requirements. Moreover, this kind of communication system can be seen as a building block of the definition of the so-called IoT and an example of M2M.

4.3.2.4 Visible light communications

Exploiting the visible radiation for communications is one of the most promising technologies for short- to medium-range next-generation nomadic access [12]. A huge available bandwidth, ultra-dense indoor channel reuse, no interference to radio channels, and perceived biological compatibility are the most relevant features. Light propagation can be spatially confined to provide secure accesses only to desired users; moreover, different wavelengths can provide multi-carrier like digital transmission schemes. Challenges to be solved are related to coexistence with other light sources, including the solar radiation, the availability of low-cost detection devices, and emission compatibility for long-range links.

4.4 New Paradigms and Challenges

4.4.1 SDN/NFV Paradigm

Individual advances on each technology components are not enough unless an efficient and versatile integration framework is built. Fortunately, the path to a seamless integration of access components and services provision has been marked in the last years through the ever increasing presence of the "virtualisation" concept [11]. NFV and the abstraction of network entities (Software Defined Networking) provide a flexible and powerful framework for the provision of next-generation digital services in a future efficient hybrid system. The physical infrastructure is distributed among different operators and technologies: the satellite/terrestrial access, the mobile networks, cloud service operators, and an IT infrastructure carrying the physical components of the NFV. The radio access itself can be virtualized producing benefits in terms of flexibility and availability (e.g. during emergency contexts) [13].

4.4.2 Cognitive Paradigm

Focusing on wireless access technologies, higher computing resources are available for the communication layers in the terminals, so complex operations can be accomplished in order to seek for appropriate and efficient spectral resources. The "cognitive" paradigm has been largely explored from the research community in the past, but it is in the next-generation networks that it will play a significant role [14]. The cognition, as in humans, is applied to solve specific problems. One of the killer challenges in modern wireless terminals is the energy impact of radio access functions. Cognitive strategies along with the application of Game Theory can realize distributed optimization of radio resources in a quickly changing environment [15].

4.4.3 Social Paradigm

The trend toward a "global connectivity" and the large availability of smart devices equipped with different sensors lead to an increase in the adoption of the social networking concepts within the future architectures.

Users may be the source of important information to be shared with others (emergency situation, intervention scene, accident, etc.). The efficient and timely sharing of information among users enables the development of innovative services for enhancing quality of life, mobility, security, and emergency management. Moreover, the concept of social interactions may be extended to "smart objects", which will autonomously establish social relationships so as to build a social network in the IoT [16].

4.4.4 HBC Paradigm

A natural evolution of current communication systems is the "remote sensorial perception". The Human Bond Communication paradigm has been recently proposed with the aim to transmit the features of a subject in the way humans perceive it [17]. The translation and transmission of the overall perception of a physical subject into the information domain allow users to access a multi-dimensional communication space. The exploitation of all five senses and their complex interactions will open to new and unexplored opportunities in the future telecommunication ecosystem.

To enable such innovative technology, several technological challenges shall be solved, ranging from the security issues to the energy efficiency ones, including electromagnetic compatibility, reconfigurability and resiliency.

4.4.5 Specific Security and Privacy Mechanisms

Fast-changing topologies and propagation conditions also have impact on security aspects of the radio communication. Delegating the security aspects to traditional layers (i.e. transport and application layers) is a weak choice in a "hyper-connected" Internet-of-Everything world. This is why physical layer security has gained recently large interest and consensus. Securing the waveforms for cognitive wireless networks is however a challenging task, but not fully addressed here [18].

4.5 Conclusions

In the future, pervasive Internet *Ecosystem,* communication, localization, and sensing capabilities will be seamless integrated for the provision of innovative services. Heterogeneous systems, smart objects, and users will be part of an overall network, as foreseen in the vision of the challenging IoE.

New architectures are necessary and rely on enabling technologies and the most promising paradigms for implementing a high flexible and reconfigurable system able to support multi services platform. 5G systems, satellite communication, satellite/terrestrial localization systems, WSNs, and visible light communications will be the main components of the future architectures and their integration will be provided through the adoption of one or more novel network concepts such as SDN/NFV, cognition, social networking, HBC, and user-controlled security.

References

[1] Regulation (EU) No 1291/2013 of the European Parliament and of the Council of 11 December 2013 establishing Horizon 2020 – the Framework Programme for Research and Innovation (2014–2020) and repealing Decision No 1982/2006/EC.

[2] Del Re, E., Morosi, S., Ronga, L. S., Jayousi, S., and Martinelli, A., Flexible Heterogeneous satellite-based architecture for enhanced quality of life applications. *IEEE Commun. Mag.* 53(5), 186–193. doi: 10.1109/MCOM.2015.7105659

[3] Del Re, E., and Morosi, S. (2013). "Flexible intelligent heterogeneous systems for enhancing quality of life," in *CONASENSE Communications, Navigation, Sensing and Services*, eds. L. P. Ligthart, and R. Prasad, pp. 43–65, ISBN: 9788792982391 (Aalborg: River Publishers).

[4] Soldani, D., Pentikousis, K., Tafazolli, R., and Franceschini, D. (2014) 5G Networks: end-to-end architecture and infrastructure. *IEEE Commun. Mag.* 52(11), 62–64.

[5] Benelli, G., Del Re, E., Fantacci, R., and Mandelli, F. (1984). "Performance of uplink random-access and downlink TDMA techniques for packet satellite networks", *Proc. IEEE* 72, 1583–1593. doi: 10.1109/PROC.1984.13055

[6] Ferrús, R., Koumaras, H., Sallent, O., Agapiou, G., Rasheed, T., Kourtis, M.-A., Boustie, C., Gélard, P., Ahmed, T. (2016). SDN/NFV-enabled satellite communications networks: opportunities, scenarios and challenges. *Phys. Commun.* 18, 95–112. doi:10.1016/j.phycom.2015.10.007

[7] Groves, P. D. (2013). *Principles of GNSS, Inertial, and Multisensor Integrated Navigation Systems*, 2nd edn. Artech House, Norwood.

[8] Morosi, S., Jayousi, S., Falletti, E., and Araniti, G. (2013). Cooperative strategies in satellite assisted emergency services. *Intl. J Satellite Commun. Network.* 31(3), 141–156. doi:10.1002/sat.1028

[9] Morosi, S., Del Re, E., and Martinelli, A. (2014) "Cooperative GPS positioning with peer-to-peer time assistance", in *Proceeding of the 4th International Conference on Wireless Communication, Vehicular Technology, Information Theory and Aerospace & Electronic Systems Technology 2014 (Wireless VITAE 2009)*, May, Aalborg, Denmark.

[10] Martinelli, A., Morosi, S., and Del Re, E. (2015). "Daily movement recognition for Dead Reckoning applications", in *Proceeding of the 2015 International Conference on Indoor Positioning and Indoor Navigation (IPIN 2015)*, Banff, Canada.

[11] Conti, A., Dardari, D., Guerra, M., Mucchi, L., and Win, M. Z. (2014). Experimental Characterization of Diversity Navigation. *IEEE Syst. J.* 8(1), 115–124.

[12] Dimitrov, S., Haas, H. (2015). *Principles of LED Light Communications: Towards Networked Li-Fi*. Cambridge University Press, Cambridge, ISBN 1316299031.

[13] Ronga, L. S., Pucci, R., and Del Re, E., (2015). Software defined radio implementation of CloudRAN GSM emergency service. *J. Signal Proces. Syst.* doi:10.1007/s11265-015-1040-2

[14] Del Re, E., Maseng, T., and Ronga, L. S. (2012). Trends and applications for cognitive radio. *J. Comput. Networks Commun.*

[15] Del Re, E., Piunti, P., Pucci, R., Ronga, L. S. (2012). Energy efficient techniques for resource allocation in cognitive networks. *J. Green Eng.* 2012329–346.

[16] Atzori, L., Iera, A., and Morabito, G. (2011). SIoT: giving a social structure to the internet of things. *IEEE Commun. Lett.*

[17] Prasad, R. (2015). "Human Bond Communication," *Wirel. Personal Commun.* doi:10.1007/s11277-015-2994-x

[18] Mucchi, L., Ronga, L. S., and Del Re, E. (2011). Physical layer cryptography and cognitive networks. *Wireless Personal Commun.* 95–109 (Netherlands: Springer).

Biographies

E. D. Re is full Professor at the University of Florence, Florence, Italy. Presently he is the Director of the Department of Information Engineering (DINFO) of the University of Florence and President of the Italian Interuniversity Consortium for Telecommunications (CNIT), having served before as Director. He is the head of the Signal Processing and Communications Laboratory of the Department of Information Engineering (DINFO) of the University of Florence.

S. Morosi received his Ph.D. degree in information and telecommunication engineering from the University of Florence in 2000. He is currently an assistant professor at the same university. His present research interests focus on communication systems, green ICT and energy-efficient wireless communications, and future wireless communication systems. He is an author of more than 100 papers in international journals and conference proceedings.

L. S. Ronga received his Ph.D. degree in telecommunications in 1998 from the University of Florence, Italy. In 1999 he joined Italian National Consortium for Telecommunications, where he is currently head of research area. His research interests span from satellite communications to cognitive radio, software defined radio, radio resource management and wireless security. He has been principal investigator in several research projects and author of more than 90 papers in international journals and conference proceedings.

S. Jayousi received her Ph.D. degree in Computer Science, Multimedia and Telecommunications in 2012 from the University of Florence, Italy. She is with the Department of Information Engineering of the University of Florence since 2008. Her research activity is mainly focused on: satellite communications for emergency, IP QoS network management in hybrid satellite/terrestrial networks, cooperative communications and diversity algorithms in relaying systems. She's author of several transaction and conference papers.

5

Organizing International ICT Research for Multidisciplinary Applications

Albena Mihovska and Ramjee Prasad

Center for TeleInfrastruktur (CTIF), Aalborg University, Aalborg, Denmark

Information and Communication Technology (ICT) has evolved over the past decade to an inherent part of daily life. The digitalization has become a backbone of many of our activities, spurring numerous new ICT-based applications in their support. Such rapid penetration of technology enabled by wireless and wired connectivity has strengthened the possibilities for humanity but at the same time has brought about new points of weaknesses due to a highly distributed character of the ICT networks and applications (i.e., security, interoperability, identification), and opening policy and legal challenges (i.e., legitimacy, transparency, accountability, anticompetitive behavior), demanding a firm focus on inter- and cross-disciplinary enabling solutions toward all-round deployable technological concepts and a protection of our rights to privacy and trust. This chapter gives an overview of the trends and evolution of ICT and analyzes the factors that demand a new approach to ICT research, both from an academic and other stakeholders' perspective.

This chapter is organized as follows. Section 5.1 gives an introduction to the topic, the evolution of ICT research, and explains the motivation for organizing ICT academic research in an international framework. Here, we also introduce the model adopted by the Center for TeleInfrastruktur (CTIF). Section 5.1 elaborates on the main research concepts related to communication, navigation, sensing, and services (CONASENSE) enabled by ICT. Section 5.2 gives the authors' perspective on the role of academia within the international ICT research. A view is given on the role of novel teaching

Role of ICT for Multi-Disciplinary Applications in 2030, 95–118.

methods, such as problem-based learning and the inclusion of standardization for stimulating research ideas of entrepreneurship value. Section 5.3 analyzes the role of standardization for promoting ICT research. Section 5.4 concludes the chapter.

5.1 Introduction

ICT research is a large field where a multitude of various specific scientific fields collaborate together in a thematic network to enable the realization of a complex application. Examples, of successful ICT applications, currently receiving a lot of research attention and, some of which have been even partially deployed, are smart grids, smart city, smart buildings, e-health, ambient-assisted living (i.e., smart home), ICT-enabled manufacturing processes, self-driving cars, and a few others. A typical research team for such an application will include engineers from various fields, social scientists, medical scientists and social workers, designers, marketing experts, and on many occasions, the individual end users of the application, which get involved by means of pilot user trials. Such trend has been observed within the European-funded research programs that allow the research to be organized under a collaborative umbrella framework not only involving a team of inter- and multidisciplinary researchers and stakeholders but also providing the means for highly international pan-European and inter-continental research [1]. The European Commission has pointed out that such an approach has the potential to strengthen the scientific and technological base of the European industry and to encourage Europe's international competitiveness, while promoting research that supports EU policies. A similar trend can be observed in national research funding programs, even the ones supporting individual post-graduate researchers. For example, one strongly present requirement in the majority of Danish funded ICT research [2] is the cooperation factor, at least between inter-disciplinary research units of the same organization or between academia and industry as an important part of the Danish innovation policy, which acknowledges that innovation is to be driven by societal challenges to a larger extent than today and that more knowledge should be translated to value.

The Center for TeleInfrastruktur (CTIF), Aalborg University (AAU) in Denmark was established on January 29, 2004 as a research and education center, with a strong focus on ICT. This was a natural development based on a strong, long-standing record of pioneering leadership in personal and wireless communication and the recognition of the trend in its future development that quickly evolved into cross-/inter- disciplinary research in a cooperation framework with all the four faculties at AAU. Accordingly, eight thematic areas were established to categorize the CTIF research, namely: Multimedia, Applications and Services, Personalized Medicine, Communication, Navigation and Sensing for Intelligent Services, Network without Borders, Resource Optimal and Embedded ICT, Cognitive Communications, and ICT for Humanities & ICT Technologies for Social Studies.

Across the years, CTIF has been involved with many industrial and academic partners through EU integrated projects. It has also received funding from many national and multinational industries. Today, CTIF has a large Global Network, which consists of academic, industrial, and other members worldwide. In addition, CTIF Global is organized in several CTIF Divisions spread on three continents. The CTIF Global Network works as a gateway to international partnerships and cooperation and provides an international platform for the purpose of strengthening the global market. CTIF strongly supports standardization and was the initiator of the International Telecommunication Union (ITU-T) initiative Standardization in Education [3].

To understand well the complexity of the ICT area, the following sections analyze the main factors for the evolution of ICT research and the effect of the ICT evolution for the emergence of novel and multidisciplinary ICT applications.

5.1.1 Evolution of ICT Research

The introduction of the cellular phone more than twenty years ago truly can be defined as the real trigger of the rapid digitalization of the present society. Over the past two decades, the society has undergone a powerful leap toward the use of computing, wireless, and Internet technologies in all spheres of the daily and business life. Step by step, we all have come to depend on the availability of the mobile device- and Internet connectivity in a similar way as we take for granted the availability of electricity and water.

Multimedia services were identified as the driving forces for radical shifts in the development of digital wireless communications ever since the emergence of non-voice mobile communications (e.g., third-generation wireless communication systems) [4–6]. This, in turn, introduced new value players and imposed a need to redefine the business models. Today, multimedia and social networks data are a predominant characteristic of the digital communication traffic, while the real driving force behind the technological advances is the user and the never-ceasing quest for faster and richer applications and ubiquitously available access to such things [7].

While in the 1990s, the computing and the wireless and fixed communications areas were evolving as parallel worlds, since the early years of the millennium, the research advances that both made on the technological and service and application side, together with the user need for seamless and highly personalized services and applications, have been forcing a merger toward what today is known as ICT.

A dominant driver of this merger has been the steady trend of convergence of disruptive key technologies as an enabler of seamless end user experiences. Figure 5.1 shows the interdependence between the main connectivity enabling technological areas to support the delivery of various services and applications enabled by the IT infrastructure.

A good overview of how communication technologies had evolved over the years since the early 1990s to deliver the mobile and wireless communication vision beyond 2010 can be found in [8]. Since then, various technologies within the communication infrastructure, such as the innovative concepts of optical and 5G communication technologies; cognitive radio, spectrum and carrier aggregation, flexible spectrum access, collaborative spectrum sharing, smart antennas and massive MIMO, high-frequency band technologies, radio frequency identification (RFID), wireless sensor networks,

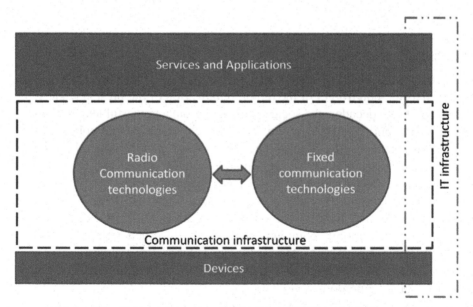

Figure 5.1   Convergence of communications and IT into the ICT concept.

and related supporting technologies, have emerged. Together with hardware advances based on concepts such as software-defined radio, nanotechnology and MEMS, and application programming interfaces (APIs), a more versatile type of portable connecting devices emerged, stimulating the end user toward using and demanding novel services and applications.

Advances within wireless sensor networks (WSNs) have allowed for large amount of data to be collected from remote sources and sent over the telecommunication infrastructure (wireless, fixed or mobile). The concept of machine-to-machine communications (M2M), together with the proliferation of Internet-connected multimedia devices, has resulted in huge amounts of unstructured data (e.g., video, click stream, log files, audio) being generated. The need to deal (i.e., process, store, etc.) with the huge amount of information, often of critical nature and demanded in real-time, has in turn given a research push toward technologies, such as mobile cloud and fog computing, server virtualization, mm-wave and visible light communications (VLC), homomorphic encryption, biometrics, context-aware services, and so forth, that somehow would need to be incorporated into a single ICT scenario. The quest for data also demanded a new approach toward resource use, both in the mobile and computing world. This gave the incentive to a concept, known today as "virtualization".

Virtualization today has found a large use for the management of resources in radio access networks as well as for data resource management. It gave a basis for ICT concepts such as the Cloud-RAN, which has great benefits for mobile network expansions [9] by allowing for wireless resources to be pooled and utilized by cooperating cells, and has recently also gained momentum for use both in macro- and small cell scenarios. In the context of C-RAN, virtualization allows for performing the baseband functions on a centralized data center IT infrastructure, while the antennas and other radio elements may remain distributed. Virtualization is an enabler of IoT, as it allows that essential features and services will work uniformly across random topology scenarios. Virtualization can be equally successfully applied for obtaining innovative ICT research solutions by allowing for sharing scientific resources and knowledge an international scale and across a cross-disciplinary expertise field, binding together various stakeholders.

The following sections, elaborate on the novel research concepts and multidisciplinary applications, which are enabled by the ICT evolution and which require research work under inter- and cross-disciplinary umbrella frameworks.

5.1.2 Influence of the ICT Evolution on the 5G Concept

The vision of 5G currently under conceptualization, widely on focus among researchers representing all stakeholders, weighs toward a two-layered structure, consisting of a radio network and a network cloud, and integrating various enablers such as small cells, multihop cellular and delay tolerant networks (DTNs), network function virtualization (NFV), and software-defined networking (SDN) [10, 11]. In addition, services are envisioned to provide intermittent or always-on hyper connectivity for machine-type communications (MTC), covering diverse services, such as connected vehicles, connected homes, moving robots, smart parking and industrial monitoring, with sensors that must be supported in an efficient and scalable manner. Convergence of technologies, ultra-high capacity, universal coverage and maximal energy, and cost efficiency are key characteristics of the 5G wireless system concept [12].

Figure 5.2 shows the envisioned scenario of a radio network infrastructure featuring partial overlap of 3G, 4G, and 5G cellular and small cell technologies. QoS-stringent machine-to-machine (M2M)/IoT/5G applications are provided by a cross-segment orchestration of PHY, MAC, and upper layer resources and exploiting NFV and SDN as key enablers for the proper resource provisioning.

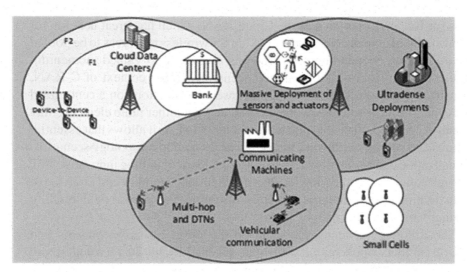

Figure 5.2   5G enabled by the convergence of novel ICT technologies.

The scenario in Figure 5.2 imposes the following limitations, namely:

- Limited available spectrum;
- Smaller and more numerous cells, increasing the need and number of low-cost, energy-efficient connections;
- Imperfect backhaul;
- Complex coexisting environment of multiple heterogeneous technologies and users.

Traffic offloading to fixed networks through Wi-Fi in unlicensed frequency bands has become wide spread as a means to deal with an overburdened licensed spectrum. Even more recently, research has commenced in the direction of VLC and high-frequency and mm-bands, as another alternative solution to deal with the rapid proliferation of unlicensed band devices and, in general, a rapid population even of the unlicensed radio bands. Such a scenario is shown in Figure 5.3.

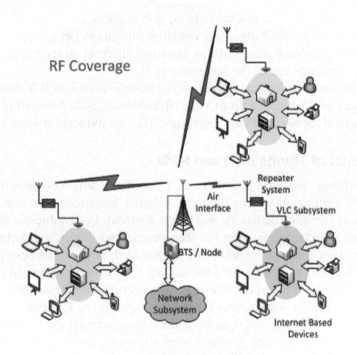

Figure 5.3 Seamless integration of VLC for radio spectrum offload [13].

The scenario of Figure 5.3 is realized in a distributive approach, where each sub-scenario benefits from the use of VLC spectrum and can be connected to the main telecommunication network via an air interface using a repeater system.

To enhance the network capability and optimize its usage, the operators are deploying more localized capacity, in the form of small cells (e.g., pico- and femto cells and remote radio units connected to centralized baseband units by optical fiber) [14]. Virtualization of wireless nodes based on software defined radio (SDR) principles, to allow them to support multiple heterogeneous transmission technologies, together with techniques for optimal resource sharing according to application requirements, and the context has also been proposed. Such scenario requires a multifold approach to tune the PHY, MAC, radio resource management (RRM), and upper layers.

Small cells coordinated and uncoordinated deployments have been extensively studied from the point of view of how to manage interference and radio resources [15, 16]. In multiuser scenarios, many users share a limited amount of resources, these resources being the medium through which the users communicate. Scheduling and resource allocation become essential components of wireless data systems because different users experience different fading conditions at the same time [17].

The realization of such scenarios is only possible within an ICT context. Concepts, such as machine learning (ML), virtualization, and cloud computing become essential for successfully operating a 5G communication scenario.

5.1.3 Internet of Things (IoT) and M2M

Internet of Things (IoT) is a prominent ICT paradigm, which enables information flows from/to and among highly distributed, heterogeneous, real, and virtual devices (sensors, actuators, and smart devices). IoT enables forms of collaboration and communication between people and things, and between things themselves, hitherto unknown and unimagined. The IoT can be defined simply as the Internet of connecting human and things. In the technical point of view, the IoT is defined as the Internet of connecting the human and things with identifiers and/or information-processing capabilities. Comparing with the existing networks, the IoT has the following significant characteristics: connecting directly with the physical world without human intervention; autonomic networking of the IoT nodes; and autonomic interaction between the IoT nodes.

The IoT is a very complex network system, as it needs to connect a variety of different types of terminals and access networks for different application purposes.

The IoT concept is probably the best example of the innovations brought about by the IT and telecommunication convergence. Most IoT applications entail a large number of heterogeneous geographically distributed sensors. As a result, they need to handle many hundreds (sometimes thousands) of sensor streams. Many IoT elements and infrastructures (e.g., sensors, WSN, RFID), however, are location specific, resource constrained, and usually expensive to be developed and deployed. Consequently, IoT infrastructures are in general inflexible in terms of resource access and availability [18, 19].

The IoT operation relies on the automatic management, identification, and use of a large number of heterogeneous physical and virtual objects (e.g., both physical and virtual representations), which are connected to the Internet. A common aspect of IoT applications is that many of them of practical interest involve control and monitoring functions, where human-in-the-loop actions are not required. As a matter of fact, a main reason for having many of these applications is to remove human intervention for improved efficiency, security, and safety.

Cloud computing and virtualization provide the means of immense distributed storage capacities that can solve the above problems. The cloud can provide the large-scale and long-lived storage and processing resources for the personalized ubiquitous applications delivered through the IoT networks as well as important backend resources. However, cloud-based platforms stay far from the real nodes connected to them. On the other hand, device-centric technologies and applications, such as IoT, constitute part of a local to the users and distributed in nature infrastructure, where a lot of personalized, and also vital, data come from sensors and actuators.

The main research challenge here is to integrate successfully the currently centralized concept of the cloud and its utilities to the highly distributed type of platforms on which IoT applications and services are based. An integrated combined framework utilizing the computing and storage capacity of the cloud to all ends of IoT communications and services can become a powerful tool to build new businesses.

M2M is a subset of the IoT, which creates a bridge between the real world (made of sensors, actuators, tags that are pervasive in our lives) and the virtual world (the Internet and its associated services).

IoT allows for the realization of many ICT scenarios (e.g., smart grid, self-driving cars, smart cities, e-Health) each having their own specific user and usage requirements. One common feature, however, is the involvement of a large number of devices in their realization. Some of those are mobile and relying on battery power for their operation. For example, even if the requirements for services delivered to e-Health or a smart home scenario may differ, common critical parameters will be security, mobility, and reliability that exist.

Typical IoT scenario characteristics are the unpredictability of the duration, location, and time of communication between IoT objects as well as the need to enable the successful completion of the communication process and error-free conveying of the information.

Despite the huge research interest in IoT, a consensus on a common IoT architecture has not been reached yet within the research community. One possible reason for that is the lack of a joint technical and legal framework where the benefits of the various proposed concepts can be weighed against the interests of all stakeholders. Inter- and cross-disciplinary research can be an enabler for establishing this basis for building the future IoT ecosystem. Research should continue to focus on fundamentally novel networking architecture concepts to support the collaborative behavior in an IoT system, including novel research on the required capabilities related to personalized service creation, identity-based user management, and IoT connectivity. Another key research aspect is to enable identification, naming, and addressing capabilities for supporting the connectivity in the end user domain. Further, through the means of joint international inter- and cross-disciplinary research, an insight can be gained on how network and environmental awareness can be perceived, developed, and expressed in an IoT scenario in order to lead to more flexible, efficient, and modularized architectural paradigms.

5.1.4 The Power of ICT

Envisioned applications of ICT include smart grids, air pollution control, monitoring the condition of crops in farming, health monitoring, reduction of hospitalization costs by means of remote health assistance, support of the quality of life of the aging population, autonomous car driving, intelligent transportation systems, controlled drug delivery, smart garbage processing to help bio-degradation, interconnected office, disaster recovery and management, e-learning, e-government, and many more.

ICT in general has the potential to solve significant social problems, such as the issues of aging, environmental problems, etc. Practically, every application could benefit from using ICT in one or another enabling configuration.

An ICT service/application refers to systems, including computers, programs, databases, people, and operational support to manage the applications. Communication within an ICT scenario refers to the information structures and networking that enable the communications between the services/applications and entities in the physical equipment. The physical equipment refers to the devices, sensors, and controllers that provide information to the ICT services/applications, and receive commands to affect the control of the devices in the physical equipment.

More and more ICT services/applications rely on the cloud-computing concept for the data management and processing. While cloud computing started with three main pillars, "Software as a Service-SaaS"; "Platform as a Service-PaaS"; and "Infrastructure as a Service-IaaS," today, the number of functionalities, that are enabled as a service, is rapidly increasing: Network as a Service (NaaS), Desktop as a Service (DaaS), Service Delivery Platform as a Service (SDPaaS), Things as a Service (TaaS) [20]. The TaaS concept was the predecessor of the concept of fog computing, namely, TaaS allowed for some cloud characteristics, such as resource pooling, rapid elasticity, and measured services to be implemented as a local content-centric cloud of gateways, providing services that map content information with thing resources, or make resources and data accessible by content regardless of physical location.

In an ICT application scenario, we need to cater for the orchestration of available data collected from different enabled sensors, collecting data and intelligent decision support system output that will correlate available data sources, environment, and ICT application context to propose the most accurate actions. To enable prompt and relevant decisions, semantic data correlations are essential in order to provide optimal situation reaction with most optimal activity-related visual data presentation [21]. Each ICT application requires an ontology, specific to its needs, that enables semantic sensor data correlation and semantic modeling of the middleware services that are distributed at the sensing and the cloud environments.

A common feature of an ICT scenario, besides the communication and information access enabled at all times, is that most of the communicated and accessed information derives its value in real time. For example, an e-Health scenario would provide a real-time feedback of the patient's vital data; a car-to-car communication scenario depends on real-time data to enable traffic safety of the vehicles and the drivers; an environmental sensing scenario would need to deliver immediate feedback in the case of fire or other disasters; a smart grid scenario would depend on real-time information to ensure network stability as required by the grid codes or to decide on how much renewable energy should be injected into the grid.

The need to process the real-time data needs communication optimization algorithms, where real and non-real time parameters information relevant to the particular scenario can be included in the optimization functions. To this end, network solutions that allow for transparent communication over the full spectrum of available communication technologies and infrastructures (e.g., Ethernet, cable, fiber, Wi-Fi, ZigBee, 3G/4G/5G . . .) are needed, hereby taking into account (1) the quality of service (QoS) constraints (latency, reliability, bandwidth . . .) dictated by the ICT applications (collection of application-related parameters, collection of individual user demands . . .); (2) the availability of wireless/wireless technologies and their policies (e.g. cost, maximum allowed bandwidth usage); and (3) the dynamic properties of the different connection technologies (bandwidth, BER, transmission delay . . .). Different communication paths may be established depending on the nature of the traffic (e.g. real-time versus non real-time, critical versus non-critical, low versus high data rate) and the instantaneous properties of the different communication links.

The concept of 'fog computing' allows for offloading the cloud computing tasks to the edge of the network, closer to the application and is another enabler of real-time data processing [22–24]. For example, in an e-Health scenario [21], which enables remote care for the patient, the one vital to the patient's medical state information will come from the sensing environment installed at the patient's home. A fog node, located closer to the sensing environment would implement real-time signal processing algorithms that would be responsible for all the real-time processing of the health-related collected data to enable a set of personalized services or that a medical staff intervenes timely in response to an alarm. An example of this scenario application is shown in Figure 5.4.

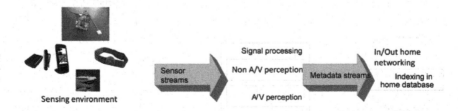

Figure 5.4   The concept of fog computing in an e-Health scenario.

ICT has the power to transform society and enable sustainability of the economy; on the other hand, the ICT services/applications raise new challenges. One important aspect that goes with ICT and digitalization is to cater for proper security solutions, especially because of the virtual and invisible to the end user in most cases handling of information, often of personal or critical nature.

5.1.5 Cybersecurity Challenges

The high number of the involved devices in an ICT scenario, most of which is mobile, the various sources of collecting data (e.g., wireless sensors), and the unpredictability of the connections to be established, define a distributed type of topology. Security is becoming an increasingly important topic of concern, both due the more distributed nature of the communications, and for the most confidential or of critical to public safety nature of the data being collected and transferred.

Cybersecurity is one major area of research currently coming into focus both from an academic, industrial, and public point of view. It is crucial for the successful deployment of ICT applications to provide a sufficient level of security spanning from confidentiality of transmission to protection against attacks (e.g., denial of service-DoS and distributed denial of service (DDoS), where an intruder aims at disrupting the operation of the network (e.g., through jamming the communication or exhausting the energy of the key data processing devices). The larger the span of the emerging ICT applications, the larger the set of possible cybersecurity threats, and consequently, the larger the magnitude of their impact. In view of the cross-disciplinarity of the ICT applications and how they touch each aspect of the society existence, a breach in cybersecurity space would have global consequences. Protection against cybersecurity crime will require that technical, legal, policy, social experts work jointly to build a solid ICT system. The challenge of cybersecurity, therefore, absolutely requires a multifaceted international research approach.

5.2 Role of Academia

The research for communication techniques for current and upcoming ICT applications and technologies has just begun in the last few years and there are countless important open problems, which is a harvest time for theoretical and practical research with break-through potential.

In view of the multifold aspects of the ICT research problematics, and also, in line with the major ICT challenge related to cybersecurity solutions, it is key that academic research adopts an inter- and cross-disciplinary nature to help unlock the hidden potential of the various ICT enabling technologies. Academia should act as a research hub for technology innovation through education based on an inter- and cross-disciplinary and international cooperation. This vision is shown in Figure 5.5.

Learning through an inter- and cross-disciplinary solving approach allows for addressing all aspects of the contemporary ICT research problem. To further strengthen the research, learning practices that encourage entrepreneurship and support commercializing of IPRs through start-ups would empower academic research staff and students to use technology solutions to promote values, ability of self-learning, and taking responsibilities.

When solving an ICT application-related problem, academic research and education should target to include experts and researchers from the relevant inter- and cross-disciplinary area around ICT, including but not limited to medical science, human and social science, business management, energy in order to build knowledge, and innovations based on upon mutual collaboration.

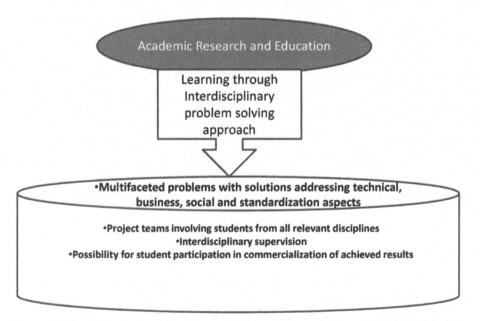

Figure 5.5   Unlocking the academic research potential through inter- and cross-disciplinary approach.

5.2.1 Problem-based Learning

Problem-based learning (PBL) is an advanced educational concept, currently gaining more interest as a way to focus research innovations around IPRs and their commercialization as well as toward techno-entrepreneurship promotion.

Education through PBL provides knowledge that is gained while trying to apply the theoretical skills to a practical problem. Such approach has been proven to lead to innovative thinking of the students as well as to teach them to solve problems both individually and in a team. PBL is the educational choice of preference at Aalborg University (AAU), Denmark [25], and provides for our academic and research staff and students opportunities to share their knowledge toward a competitive solution for an ICT challenge. Further details on the PBL method can be found in [26].

5.3 ICT and Standardization

ICT applications act across a whole range of functions that have an impact on a particular sector of the society.

ICT systems include tools for public or private authorities and professionals, from national to international, data-processing specialists, social security administrators, and—of course—the end users, both the individual ones and the community. A good example may be given with the previously touched e-Health scenario, which includes health information networks, electronic health records, telemedicine services, personal wearable, and portable communicable systems including those for medical implants, health portals, and many other ICT-based tools assisting disease prevention, diagnosis, treatment, health monitoring, and lifestyle management. In order for ICT to prove its sustainability, a solution of where disparate domains are intelligently cooperating is required. One of the major problems identified is the lack of ICT standards, especially for interoperability; however, the latter is a key to gluing the various ICT elements together. ICT environments are extremely complex and challenging to manage, as they are required to cope with an assortment of users and business conditions under various circumstances and with a number of resource constraints. Reaching a consensus on common standards requires that experts from all fields work together to acknowledge the place in the design process of the respective critical requirements.

Interoperability testing is one such approach, which has become more and more attractive in standardization and industry to ensure delivery of services across products from different vendors. It has been recommended that testing should span at least over:

- Platform testing;
- Conformance testing;
- Application and content testing;
- Field and interoperability testing;
- Operator pre-acceptance/acceptance testing;
- Performance, localization, usability testing, and certification.

An information and communication system is fundamental to achieving intelligent management and control in any ICT scenario. It builds up a multiway information channel to achieve interaction (e.g., demand response, real-time price, etc., which are typical for a smart grid scenario). Usually, it is also needed to implement solutions for auto-collecting and analyzing of the obtained information. In order to ensure networked communication and interoperability between the various ICT scenario elements/devices, a unified, open communication infrastructure is the key. This is also a way to strengthen the robustness and ability of self-healing of the ICT system supporting a particular scenario.

The benefits of ICT standards are summarized in Figure 5.6.

The lack of well-established standards limits the potential of the ICT solutions to bring in the full personal and economical benefits to society, and, therefore, needs to be addressed not only by the industrial stakeholders but also by the academic education and research. However, standardization is a

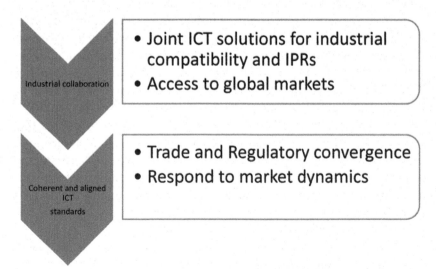

Figure 5.6   Benefits of ICT standards.

key to take into account the mind gaps between the various types of industries comprising an ICT scenario (e.g., telecommunication and power for a smart grid; or telecommunication and medical for an e-Health scenario). The role of standardization is to clarify the existing ICT-related activities of standards bodies and to obtain requirements requested by the respective ICT industries.

Major standardization players toward conforming ICT standards are bodies such as the International Telecommunication Union (ITU) [3], the European Telecommunication Standards Institute (ETSI) [27] and the Global ICT Standardisation Forum for India (GISFI) [28], among others.

5.3.1 Academia and Standardization

Standardization is the key for an economic growth in a global context—both for developed and developing countries; and is a strong tool for creating a green economy. It can release new potential within education with new technologies. The need to address international standardization in ICT in the academic curricula is vital for the students of today because they are the experts who will drive the standardization processes of tomorrow.

Collaborative academic research, as explained in Section 5.2.1, has the potential to create a hub for technology innovation through higher education. Therefore, knowledge about entrepreneurship and standardization are important tools to lead toward commercially feasible academic research innovations. Incorporating entrepreneurial and standardization education in the technical academic curricula will initiate patent-based research in ICT, which in turn will create a platform for future expanded collaboration with industries. This scenario is shown in Figure 5.7.

Recently, an initiative toward the inclusion of standardization in academic curricula was made by ITU-T and the Center for TeleInfrastruktur (CTIF) [29]. In the context of this joint initiative, "standards education" is related not to technology topics, but rather to the activities that aim at providing formal information to students at undergraduate and graduate levels on all aspects related to international standards, standards development activities, standard strategy planning, and business case studies using standards. This information includes inter alia importance of standardization, strategies for standardization, case studies, etc. Given the scale of the task, it is important to include experts in standardization, representatives of academia, as well as other standard development organizations that are also interested in standard education.

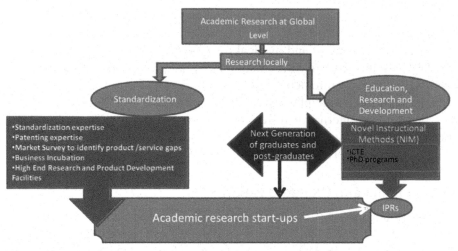

Figure 5.7 Importance of entrepreneurship and standardization for academic research innovations.

5.4 Conclusions

ICT builds upon the convergence of telecommunications and IT infrastructures and paves the way toward better and cost-effective living. ICT research problems, however, have a complex nature and require a collaborative approach toward commercially viable solutions to allow for a careful study and analysis of the technical principles, powers and market mechanisms involved. Research capacity and network building are vital tools for in order to develop a sustainable and proactive scientific environment enabling breakthrough ICT innovations. There is a need for a critical mass and a multidisciplinary approach to research and education, which should also involve standardization closely, as ICT builds upon a coherence of the technical and business requirements of the various end users. The challenges remaining for the successful organization of international ICT research problems and scenarios are closely related to adopting research frameworks that clearly recognize the added value of bringing several disciplines together and are designed to extract the full benefits of this added value. ICT problems require that several stakeholders would engage in the research, development, deployment, operation, and use of the respective research solution to also provide it with high innovation and marketing value. Therefore, research should convene on both a local and global scale to allow for the various ICT business chain players to be involved in the end solution. The success of ICT will depend on the ability to build an ecosystem supported by the appropriate regulatory environment, and a climate of trust, where critical issues such as identification, privacy, security, and interoperability are pivotal.

References

[1] Press Release-HORIZON 2020. Available at: 27http://eeas.europa.eu/del egations/mozambique/press_corner/all_news/news/2015/20151014_01_e n.htm

[2] Danish Council for Strategic Research. Available at: http://ufm.dk/en/res earch-and-innovation/councils-and-commissions/former-councils-and-commissions/the-danish-council-for-strategic-research

[3] International Telecommunication Union (ITU). Available at: www.itu.int

[4] Mihovska, A., Pereira, J., and Prasad, R. (2001). "Wireless Multimedia: Trends and Requirements," in *Proceedings of IEEE VTC'01 Spring*, Rhodos, Greece.

[5] Prasad, R., (Ed.) (2001). *Towards a Global 3G System*, Series. Artech House, Norwood.

[6] Prasad, R., Mohr, W., and Konhäuser, W. (2000). *Third Generation Mobile Communication Systems*. Artech House, Norwood.

[7] Mihovska, A., et al. (2015). "Integration of wireless and data technologies for personalized smart applications," in *Proceedings of Wireless Telecommunications Symposium (WTS)*, April, 2015, (New York, NY. IEEE Press).

[8] Mihovska, A., et al. (2007). "Towards the Wireless 2010 Vision: A Technology Roadmap," *Int. J. Wirel. Per. Commun.* 42, 303–336.

[9] Fujitsu Report. "The benefits of the C-RAN for mobile network expansion." Available at: https://www.fujitsu.com/us/Images/CloudRAN wp.pdf

[10] Horizon 2020 Advanced 5G Network Infrastructure for Future Internet PPP—Creating a Smart Network that is Flexible, Robust and Cost Effective. Available at: http://ec.europa.eu/digital agenda/futurium/en/content/ horizon-2020-advanced-5g-network-infrastructure-future-internet-ppp-c reating-smart-network (Accessed October 2014).

[11] Akyildiz, I.F., et al. (2014). LTE-Advanced and the evolution to Beyond 4G (B4G) systems. *Phys. Commun.* 10, 31–60.

[12] Prasad, R., and Mihovska, A. (2013). *ITU News, No. 10.* Available at: https://itunews.itu.int/en/4619-Challenges-to-5G-standardization.no te.aspx, December 2013.

[13] Kumar, A., Mihovska, A., Kyriazakos, S., and Prasad, R. (2014). Visible light communications (VLC) for ambient assisted living. *J. Wirel. Personal Commun.* 78, 1699–1717.

[14] Hoydis, J., Kobayashi, M., and Debbah, M. (2011). Green Small cell Networks: a cost- and energy-efficient way of meeting the future traffic demands. *IEEE Vehic Technol. Mag.* 37–43.

[15] Semov, P., Mihovska, A., Poulkov, V., and Prasad, R. (2014). "Increasing throughput and fairness for users in heterogeneous semi coordinated deployments," in *Proceedings of the IEEE WCNC 2014*, Istanbul, Turkey, April.

[16] Semov, P., Mihovska, A., Poulkov, V., and Prasad, R. (2015). "Self-resource allocation and scheduling challenges for heterogeneous networks deployment," in *Intl. J. Wireless Personal Commun.* 1–19. doi:10.1007/s11277-015-2640-7

[17] Craciunescu, R., Halunga, S., and Fratu, O. (2012). "Guard interval effects on OFDM/BPSK transmissions over fading channels," in *Telecommunications Forum (TELFOR)*, 20th, Nov., 471–474.

[18] Lee, K., and Hughes, D. (2010). "System architecture directions for tangible cloud computing," in *International Workshop on Information Security and Applications (IWISA 2010)*, Qinhuangdao, China, October 22–25.

[19] Lee, K. (2010). "Extending sensor networks into the cloud using amazon web services," in *IEEE International Conference on Networked Embedded Systems for Enterprise Applications*.

[20] EU FP7 ICT project BeTaaS–*Building the environment for the Things as a Service*. Available at: http://www.betaas.eu/

[21] EU FP7 ICT Project eWALL. Available at: url://ewallproject.eu

[22] CISCO: Visual Networking Index Global Mobile Data Traffic Forecast 2014–2019, CISCO whitepaper, (2015).

[23] Stojmenovic, I. (2014). "Fog computing: A cloud to the ground support for smart things and machine-to-machine networks," *Telecommunication Networks and Applications Conference (ATNAC), 2014 Australasian*, pp. 117–122.

[24] Craciunescu, R., Mihovska, A., Mihaylov, M., Kyriazakos, S., Halunga, S., Prasad, R. (2015). "Implementation of fog computing for reliable e-health applications," in *Proceedings of the IEEE ASILOMAR Conference*, Pacific Grove, CA. USA.

[25] Aalborg University (AAU), Denmark. Available at: url://aau.dk

[26] Problem-based learning (PBL) at AAU. Available at: http://www.en. aau.dk/about-aau/aalborg-model-problem-based-learning
[27] European Telecommunication Standards Institute (ETSI). Available at: url://etsi.org
[28] Global ICT Standardisation Forum India (GISFI). Available at: url://gi sfi.org Center for TeleInfratruktur (CTIF). Available at: url://ctif.aau.dk

Biographies

A. Mihovska is currently involved in research related to defining innovative applications and research concepts for 5G communication systems, such as the integration of radio and visible light communication networks as a new 5G infrastructure, the design and implementation of e-Health services and optimization, and support of reliable and high-performance intensive data rate communications as required by the Internet of Things. She is a lecturer and supervisor in the Master and PhD study program at the Center for TeleInfrastruktur (CTIF), Aalborg University. She is involved with ITU-T and ETSI standardization in the area of Internet of Things and Smart Body Area Networks.

Prof. R. Prasad has been holding the Professorial Chair of Wireless Information and Multimedia Communications at Aalborg University, Denmark (AAU), since June 1999. Since 2004, he is the Founding Director of the Center for TeleInfrastruktur (CTIF-http://www.ctifgroup.dk/), established as a large cross-/multi-disciplinary research center at the premises of Aalborg University. Under the leadership of Ramjee Prasad, CTIF has emerged as a prominent international center of excellence for his visionary ideas and path-breaking research in wireless communications. CTIF boasts of a global presence today with its divisions dotted in eleven countries (and counting) spanning across three continents. As well as, through the numerous valuable partnerships, it has forged with world renowned academic institutions spread across fourteen countries and six continents.

He is a Fellow of the Institute of Electrical and Electronic Engineers (IEEE), USA, the IET, UK, the IETE, India, and the Wireless World Research Forum (WWRF), and is a member of the Netherlands Electronics and Radio Society (NERG), and the Danish Engineering Society (IDA).

For his exceptional contribution to the internationalization of the Danish telecommunication research and education, in 2010, Ramjee Prasad was awarded the Knight of the Order of Dannebrog (Ridderkorsetaf Dannebrogordenen-2010) by the Queen of Denmark.

Ramjee Prasad is the recipient of many international academic, industrial, and governmental awards and distinctions. He has received many prestigious international awards such as: IEEE Communications Society Wireless Communications Technical Committee Recognition Award in 2003 for making contribution in the field of "Personal, Wireless and Mobile Systems and Networks", Telenor's Research Award in 2005 for impressive merits, both academic and organizational within the field of wireless and personal communication, 2014 IEEE AESS Outstanding Organizational Leadership Award for: "Organizational Leadership in developing and globalizing the CTIF (Center for TeleInfrastruktur) Research Network", and so on.

He received an honorable award with a Gold Medal in 2014 from the Academic Council of Technical University-Sofia, Bulgaria, for his major contribution to the development of its international cooperation.

He has published more than 30 books, 1000 plus journals and conferences publications, more than 15 patents, and has guided over 100 PhD Graduates and larger number of Masters (over 250). Several of his students are today worldwide telecommunication leaders themselves. His research publications have been very well cited globally.

6

Convergence of Secure Vehicular *Ad Hoc* Network and Cloud in Internet of Things

Nandkumar Kulkarni[1], Neeli Rashmi Prasad[1], Tao Lin[2],
Mahbubul Alam[2] and Ramjee Prasad[1]

[1]Center for TeleInFrastruktur, Aalborg University, Aalborg, Denmark
[2]Movimento Group, Sunnyvale, CA 94085, USA
E-mail: {npk, np, prasad}@es.aau.dk;
{tao.lin, mahbubul.alam}@movimentogroup.com

Vehicular *Ad hoc* Network (VANET) is a highly mobile autonomous and self-organizing network of vehicles. VANET is a particular case of Mobile *Ad hoc* Network (MANET). With the recent advances in the arena of Information and Communication Technology (ICT) and computing, the researchers have envisioned that VANET could be the basis of many new applications in the field of Internet of Things (IoT). The applications of VANET are not limited to driver safety, traffic management, entertainment, commerce, etc. In the future, VANET is expected to transport the enormous amount of information. Some of the challenges in VANET are lesser computing capability, smaller onboard storage, safety, reliability, etc. Among the number of solutions proposed recently, Vehicular Cloud Computing (VCC) is one of them. VCC is a technology that provides on-demand services namely Software-as-a-Service (SaaS), Storage-as-a-Service (STaaS), Platform-as-a-Service (PaaS), etc., over the Internet via Cloud vendors.

6.1 Overview

6.1.1 Introduction to VANET

Alerting the drivers about the road situations and improving road safety is a subject of deep interest and research. Today, more and more people are affording buying cars as the car manufacturers are manufacturing more and

Role of ICT for Multi-Disciplinary Applications in 2030, 119–148.

more sophisticated cars at a reduced prize. Although human error is the prevailing cause of collisions, creators of technologies used in vehicles have an obvious vested interest in helping lower the distressing statistics. Pedestrian deaths rose by 3.1% in 2014 according to the National Highway Traffic Safety Administration's Fatal Analysis Reporting System (FARS). In that year, 726 cyclists and 4,884 pedestrians were killed in motor vehicle crashes. And this damage to innocent bystanders does not include the growing death rate of drivers and their passengers themselves [33].

Distracted driving accounted for 10% of all crash fatalities, killing 3,179 people in 2014 while drowsy driving accounted for 2.6% of all crash fatalities, killing 846 people in 2014. The road carnage is hardly limited to the United States. The International Organization for Road Accident Prevention [34] noted a few years ago that 1.3 million road deaths occur worldwide annually and more than 50 million people are seriously injured. There are 3,500 deaths a day or 150 every hour and nearly three people get killed on the road every minute.

Accidents have been accepted as serious problems, and a significant challenge to the modern society [1, 2]. With the recent evolvement in the field of Information and Communication Technologies (ICT), networks those are suitable for safety applications that communicate through wireless networks are envisioned. The prime goal of such network is to avert roadside accidents and passenger safety. For the last few years, one such network that has received more attention by most of the researchers in the field of ICT is Vehicular Ad Hoc Network (VANET) [3]. VANET is a new technology that incorporates wireless networks into moving vehicles using a Dedicated Short Range Communication (DSRC). DSRC is fundamentally IEEE 802.11a standard revised for 802.11p [4]. With VANET, vehicles on the road can communicate with other vehicles directly forming Vehicle-to-Vehicle communication (V2V). The vehicles can communicate with fixed equipment next to the road, referred to as Road Side Unit (RSU) creating Vehicle-to-Infrastructure communication (V2I) [5].

Once deployed, the advantages of Vehicle-to-Everything (V2X) are extensive, alerting drivers to road hazards, the approach of emergency vehicles, pedestrians, or cyclists, changing lights, traffic jams, and more. The technology can control car systems like brakes and power to help reduce possible bad outcomes while this new area for car technology underscores a coming time when cars can talk to each other and to road sensors.

About the only problem with V2X is that it is emerging as a perplexing stew of acronyms such as V2V, V2I, V2D, V2H, V2G, and V2P that require some explanation and the technology, while important, is not quite here yet. But the significance of this technology is undeniable so getting proficient in understanding V2X is valuable in tracking future vehicle features that will link cars to the world around them and make driving safer in the process (Figure 6.1).

6.1.2 Characteristics of VANETs

VANETs have numerous properties that distinguish it from other Ad Hoc wireless networks. Vehicles (Nodes) in VANETs are highly mobile. Owing to high mobility, the probability of partitioning the network is very high [6, 4]. End-to-End connectivity is not guaranteed. VANETs have dynamic topology, but the movement of the vehicles is predictable, and they are not completely random. The vehicles are restricted to the roads on which they are traveling. This restriction has the advantage of selecting a best route from source to destination. This predictable behavior of the vehicles also has an advantage during the link selection. VANETs are scalable. In VANETs, the bandwidth issue also intensified due to crossings, traffic bottlenecks, and the existence of buildings beside the roads, particularly in an urban environment [7].

6.2 Cloud Computing

With the rapid development of processing, storage technologies, and the success of the Internet, Cloud Computing (CC) has become a central area of investigation in the global industry. The National Institute of Standards and Technology (NIST) defines cloud computing as: "Cloud computing is a model for enabling convenient, on-demand network access to a shared pool of configurable computing resources (e.g., networks, servers, storage, applications, and services) that can be rapidly provisioned and released with minimal management effort or service provider interaction" [8].

Today, computing resources have become inexpensive. They are more powerful and universally available for computing than ever before. This technological inclination has allowed infrastructure providers to let out resources (e.g., CPU and storage) to service providers. The service providers rent and relinquish the resources on demand basis through the Internet to serve the end users. The infrastructure providers manage cloud platforms and lease resources according to a usage-based pricing model.

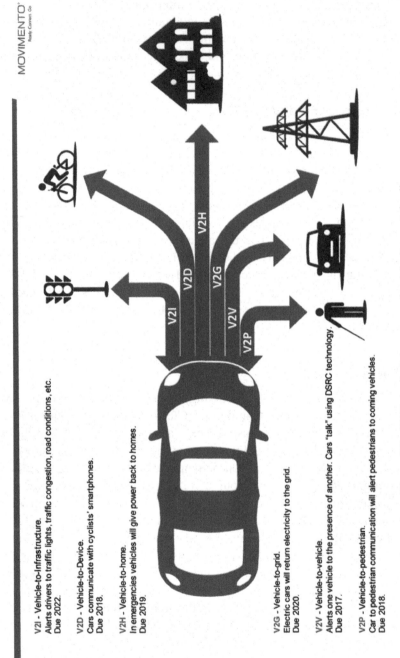

MOVIMENTO®
Ready Connect Go

V2I - Vehicle-to-Infrastructure.
Alerts drivers to traffic lights, traffic congestion, road conditions, etc.
Due 2022.

V2D - Vehicle-to-Device.
Cars communicate with cyclists' smartphones.
Due 2018.

V2H - Vehicle-to-home.
In emergencies vehicles will give power back to homes.
Due 2019.

V2G - Vehicle-to-grid.
Electric cars will return electricity to the grid.
Due 2020.

V2V - Vehicle-to-vehicle.
Alerts one vehicle to the presence of another. Cars "talk" using DSRC technology.
Due 2017.

V2P - Vehicle-to-pedestrian.
Car to pedestrian communication will alert pedestrians to coming vehicles.
Due 2018.

Figure 6.1   V2X terminology, use case, and deployment timeline.

The advances in CC have made a remarkable impact on the Information Communication Technology (ICT) industry over the last decade. Many organizations produce a large amount of sensitive data. Storage of that data on local server or machine results into the maximization of storage cost and maintenance cost. Most of the business organizations have changed their business models, and they have opted to store such vast data on a cloud in order to earn profit from this new paradigm. Companies such as Google, Amazon, and Microsoft have started providing more robust, reliable, and cost-efficient cloud platforms. CC provides several convincing features that have attracted the business, which are listed below.

6.2.1 Advantages of Cloud Computing

- **Highly scalable:** A service provider can quickly expand its service to a large extent as the Infrastructure providers pool vast amount of resources from data centers and make them easily accessible.
- **Easy access:** Any computing devices like mobile, laptop, desktop, etc., which has Internet capability, can easily avail services hosted in the cloud, as they are web based.
- **No investment:** CC uses a pay-as-you-go pricing model. A service provider need not invest in the infrastructure. The service provider simply leases resources from the infrastructure providers according to its needs and pay for the usage [9, 10].
- **Reduction in the operating cost:** The service provider can acquire and relinquish resources on demand basis. They need not provision resources in advance for the peak load. This strategy will reduce the operating cost significantly when the service request from the end user is little.
- **Reducing business threats and maintenance expenditures:** A service provider swings its business risk (such as hardware failures) to infrastructure providers by subcontracting the cloud services. In addition, a service provider can cut down the equipment maintenance and the staff training costs.

6.2.2 Cloud Computing Architecture

The design of layered architecture in the cloud is similar to the design of layers in OSI model. The cloud architecture defines four Layers that describe how applications running on network-aware user devices may communicate with the cloud. The four layers are the hardware layer, the infrastructure layer, the platform layer, and the application layer, as shown in Figure 6.2.

Layer 1 is the Hardware Layer (HL). This layer is accountable for handling the physical resources of the cloud, including servers, routers, switches, power, and cooling systems. In practice, the hardware layer is part of data centers. A data center is a collection of many servers. These servers are usually mounted in racks. They are interconnected through switches, routers, or other fabrics. Typical issues at hardware layer include configuration of the physical hardware, fault tolerance, traffic administration, energy conservation, and resource organization.

Layer 2 is the Infrastructure Layer (IL) or Virtualization Layer (VL). The infrastructure layer partitions the physical resource using virtualization technique and it generates a pool of storage and computing resources. The dynamic resource allocation is the primary task of the infrastructure layer.

Layer 3 is the Platform Layer (PL) that is built on top of the infrastructure layer. The platform layer comprises operating systems and application frameworks. The aim of the platform layer is to reduce the burden of deploying applications directly into VM containers.

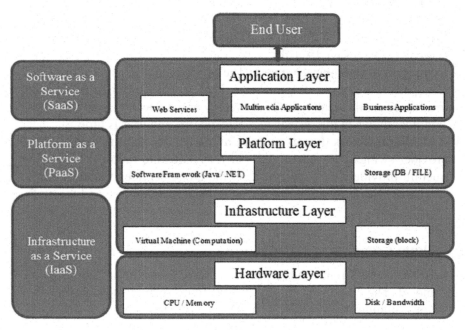

Figure 6.2   Cloud computing architecture.

Layer 4 is the Application Layer (AL). AL is the topmost layer in the hierarchy, and it comprises actual cloud applications. Cloud applications have an automatic scaling feature in order to achieve better performance, availability, and lower operating cost as compared with traditional applications.

6.2.3 Cloud Business Model

Cloud computing (CC) has emerged as a new paradigm for accommodating and providing services over the Internet. In the recent years, CC has fascinated many business owners as it removes the need for resource reservations in advance. It allows business enterprises to start with small resources when the service demand is low and increases additional resources only when there is a rise in demand. Figure 6.3 depicts the business model of cloud computing. In the layered architecture described in the previous section, every lower layer provides some service to the upper layer. In fact, every lower layer can be implemented as a service. Clouds offer services that can cluster into three categories [10]

- Software as a Service (SaaS) denotes providing on-demand applications over the Internet. An example of SaaS services includes applications related to analytical, browsing, interactive, transaction, etc.
- Platform as a Service (PaaS) means to provide resources like OS, DBMS, and Software development framework.
- Infrastructure as a Service (IaaS) talks about the on-demand provisioning of infrastructural resources, usually in terms of VMs. Multiple types of virtualization occur in this layer.

Figure 6.3 Cloud computing business model.

It is perfectly all right that the PaaS provider runs its cloud on top of an IaaS provider's cloud. However, today, IaaS and PaaS providers are often parts of the same enterprise. That is why PaaS and IaaS service providers are often called the infrastructure providers or cloud providers.

6.3 Vehicular Cloud Computing

The CC model has enabled the utilization of surplus computing power. Recently, Olariu [5] has coined the concept of Vehicular Cloud (VC). The definition of VC is as follows:

"A group of mostly autonomous vehicles whose corporate computing, sensing, communication and physical resources can be coordinated and dynamically allocated to authorized users."

The idea in VC is the large number of vehicles on highways, street ways, and parking lots that can provide public services. Each vehicle on the road or in the parking lot will be treated as useful, under utilized computational resources, and each vehicle is acting as a node in the cloud. The vehicle owners of parked vehicles in the parking lot and the owners of the cars stuck in the traffic jam can contribute to the computing power of parking garage or city traffic data centers. In these cases, the vehicles can help the local authorities in improving parking conditions or traffic conditions on the roads by sending correct messages in a timely fashion (Figure 6.4). The new VC can help in resolving serious and unexpected problems by utilizing the computing power of self-organized autonomous nodes (vehicles) dynamically [11].

IEEE has amended the basic Wi-Fi IEEE 802.11 standard in order to add Wireless Access in Vehicular Environments (WAVE) to support specificities of Intelligent Transportation Systems (ITS) applications. The new standard is called 802.11p. A spectrum band of 75 MHz is allocated in the range of 5850–5925 MHz. This allocated spectrum can be utilized for priority road safety applications, Vehicle-to-Vehicle (V2V) Communication called (inter-vehicle), and infrastructure communication called Vehicle to Infrastructure (V2I) [11–13] (Table 6.1).

6.4 Convergence of VANET and CLOUD

The idea in VANET is the large number of vehicles on highways, street ways, and parking lots that can provide public services. With VANET, vehicles on the road can communicate with other vehicles directly forming V2V

Figure 6.4 Convergence of vehicle and cloud computing.

communication. The vehicles can communicate with fixed equipment next to the road, referred to as Road Side Unit (RSU) creating V2I [14–16]. IEEE has amended the basic Wi-Fi IEEE 802.11 standard in order to add Wireless Access in Vehicular Environments (WAVE) to support specificities of Intelligent Transportation Systems (ITS) applications. The new standard is called 802.11p [17–19]. A spectrum band of 75 MHz is allocated in the range of 5850–5925 MHz. This allocated spectrum can be utilized for priority applications comprising V2V and V2I communication.

6.4.1 Data Compression

Today, more and more people are affording buying cars as the car manufacturers are manufacturing more and more sophisticated cars at a reduced prize. VANET could be the foundation of numerous new applications in the field of IoT. In future, VANETs are expected to carry the gigantic information.

Table 6.1　Taxonomy of vehicular cloud computing

Vehicular Cloud Computing	Vehicle to ReaL (V2RL)	Vehicle to Vehicle (V2V)	Vehicle to Infrastructure (V2I)	Vehicle to User (V2U)
Basic Blocks	• Emergency Vehicle Detection Basic • Crash Warning Sign Detection • Road Condition Warning • Emergency Breaking • Safely Distance Detection • Lane Change Warning • Overtaking Vehicle Warning	• Propagation of Warning Detection • Propagation of Lane Change • Propagation of traffic light status • Warning Blind Spots • Propagation of Speed Limit • Propagation of Distance Co-Ordination • Information of Turn and Direction	• Pedestrian Detection • Propagation Information Wireless • Dangerous Zone Detection • Detection Traffic Light • Communication with Pedestrian • Detection Bicycle	• Personalized Navigation • Call via Smartphone • Share Events • Connect with Home • Connection with social content • Tourism
Projects/ Applications	• Honda: Traffic Congestion Minimizer • Pioneer: 'Cyber Navi' or 'AVIC-ZH99HUD.'	• Intelligent Truck Parking Application • Vision of Internet of Cars • An independent traffic cautioning system with Car-to-X communication		• Nokia: Connected Car Fund • W3C launched work on Web Automotive • Toyota: 'Toyota Friend' • Alcatel-Lucent and Toyota: LTE connected car

- BMW: Head-Up Display
- Intel: Augmented Reality and In-vehicle Infotainment System
- PSA Peugeot Citroën: 'Peugeot connect Apps'
- BMW: 'XL Journey Mate Mini'
- Land Rover: 'Concept Discovery Vision'
- SystemX: Localisation in Augmented Reality
- Google: Driverless Car
- Volvo: Magnet Project
- Ford: 'Blueprint for Mobility.'

- Volvo Car: SARTRE
- Audi: 'Travolution project.'
- Swarco
- Volkswagen: 'C3World, connected car in a connected world.'
- Denso Corporation: Field Operational Test
- SmartWay
- US Department of transportation: Connected Vehicle Safety Pilot Program
- Compass4D
- SCORE@F in France

- Mercedes: 'Fleet tweet'
- Ford: 'Facebook and Ford SYNC Hackathon.'
- Microsoft and West Cost Custom: 'project Detroit.'
- Mercedes: Mbrace2
- Bosch and University of St. Gallen: IoTS Lab
- Toyota: 'Windows to the world.'
- General Motor: 'Windows of opportunity.'
- Audi: Future Urban Personal Mobility
- BMW: Connected Drive Concept Car
- Mercedes: 'DICE' (Dynamic Intuitive Control Experience)
- Valeo Project

To access information on any occasion, by anyone at any place, Cloud Computing (CC) is the solution. The cloud service providers manage cloud platforms and lease resources according to a usage-based pricing model [20, 21]. In STaaS, to benefit the end user in terms of price storage overhead should be as minimum as possible. Data compression techniques can be used to reduce the storage overhead on cloud. This reduction improves transmission efficiency with saving in the channel bandwidth requirement. Data compression becomes particularly important in the transmission of multimedia such as text, audio, and video. There are numerous data compression algorithms available for lossy and lossless compression. If the data are compressed with lossy data compression techniques, then entire data cannot be recovered so it is beneficial to use lossless technique to recover the original data. This section discusses compression methods that are modern one. Today, these methods are mostly of interest, since they are efficient and can compete with the modern compression methods developed through the past numerous years. Different lossless data compression algorithms discussed are Brotli, Deflate, Zopfli, LZMA/LZMA2, and BZip2. A comparative analysis of these techniques is performed, which will help the researchers in this field.

Brotli [22] is an amalgamation of modified LZ77 algorithm, Huffman coding, and second-order context modeling. It is a general-purpose lossless compression algorithm that compresses data using a very high compression ratio. It is equivalent to the best presently available general-purpose compression techniques. The performance of brotli compression algorithm is comparable with deflate on speed but, brotli has more compression ratio (nearly 10–15% more than Deflate). Brotli is free under the Apache License, Version 2.0. Brotli is very fast in compression than zopfli, and It provides 20–26% higher compression ratio. Brotli has a whole new data format. Brotli is as fast as Deflate implementation but, it compresses slightly more than LZMA and bzip2. The higher compression ratio is achieved due to second-order modeling, re-cycling of entropy codes, the larger window size for the past data, and joint distribution codes. The high compression ratio of Brotli allows smaller storage overhead and faster page loads. The smaller output size will help vehicles and cell phone users in reducing battery usage and data transfer costs. The disadvantage of Brotli is that it is not supported by existing systems (e.g. many browsers).

Deflate [23] is a blend of the Huffman coding and LZ77 algorithm [9]. It is an example of the lossless compression algorithm. The Huffman trees for each block are independent of those for previous or subsequent blocks. The LZ77 algorithm may use a reference to a duplicated string occurring in a previous block. The efficiency of Deflate is comparable with the best general-purpose compression techniques available in the market. Deflate compression algorithm works well even for a randomly long input sequence at the cost of intermediate storage. The advantage of Deflate algorithm is that it is autonomous with reference to CPU type, OS, file system, and character set. It is freely available and can be implemented readily. Deflate is harmonious with gzip utility file formats.

Lempel–Ziv–Markov (LZMA) [24] chain algorithm is a dictionary-based lossless data compression algorithm. It uses complex dictionary data structures, and a dynamic programming algorithm for encoding one bit at a time. It was first introduced in the 7z format of the 7-Zip archiver. LZMA maintains decompression speed similar to other commonly used compression algorithms. LZMA uses a dictionary compression scheme similar to LZ77. The compressed output is then encoded with a range encoder, using a complex probability prediction model of each bit. Preceding to LZMA, most encoder models were purely byte based. The novel idea LZMA brought is a generic bit-based model. LZMA gives much better compression ratio because it avoids mixing unrelated bits together in the same context. LZMA has a higher compression ratio than bzip2. It has a variable compression-dictionary size. LZMA2 is like a vessel that holds both the uncompressed and LZMA-compressed data. It also provisions multi-threaded compression and decompression. It has the ability to compress data that are not compressible with other compression algorithms.

Zopfli [25] compression algorithm is significantly slower in terms of compression speed, but it is the most size efficient deflate variant. Zopfli compression algorithm's data format is compatible with data formats of Deflate, gzip, and zlib. Zopfli can produce raw Deflate data stream or compressed data in gzip or zlib formats. Zopfli algorithm was released by Google under the Apache License, Version 2.0. Zopfli Compression Algorithm was created by Lode Vandevenne and Jyrki Alakuijala, based on an algorithm by Jyrki Alakuijala. Owing to considerably slow speed during compression, zopfli is made less suitable for on-the-fly compression. It is mainly suitable where static compression is required. The Zopfli compression method is based on iterative entropy modeling and a shortest path algorithm in order to find a low bit cost path through the entire graph. The Zopfli algorithm can be used to compress Portable Network Graphics (PNG) files as PNG uses a DEFLATE compression layer.

BZip2 [26] is open-source, patent-free high-quality data compression algorithm. BZip2 has a compression gain of typically 10–15% over the available techniques. BZip2 helps in recovering media errors. The commands to use Bzip2 are similar to GNU Gzip. It is portable and runs well on 32-bit as well as 64-bit machines with unix or win32 machines. As compared with Brotli, Deflate, and Zopfli BZip2 is very slow in compression and decompression.

The comparison of different compression techniques is illustrated in Table 6.2.

6.4.2 Assumptions and Network Model

Figure 6.5 shows the proposed model of convergence of VANET and cloud. Here, all vehicles that are part of Intelligent Transport System (ITS) are equipped with OnBoard Units (OBUs). Using wireless communication OBUs are capable of communicating with OBUs of another vehicle termed as Vehicle-to-Vehicle Communication (V2V). OBUs can also communicate with Road Side Units (RSUs) called as Vehicle-to-Infrastructure communication (V2I) [27–30]. The RSUs are a wireless static access points mounted on the road transport network that supports information exchange with OBUs. RSUs act like gateways for vehicles to the cloud services. High-speed wired Internet connection could be used for the data transfer from RSUs to the cloud. All the vehicles are Internet enabled. The vehicles can form a static Vehicular Cloud (VC) for sharing their computing resources. Vehicles parked in the parking lot of a big organization or vehicles stuck in the traffic jam on the highway can form VC. On the other hand, the dynamic cloud can be formed on demand. Vehicles using clouds [31] can connect the vehicles to the traditional clouds. Vehicles with Internet access can offer Network as a Service (NaaS) for other vehicles on the road if they need the net access. Some vehicles are equipped with higher onboard storage capacity. If one of the vehicles requires storage space for the execution of its applications, then the vehicle having larger storage capacity can provide storage as a service (STaaS) [32].

6.4.3 Block Schematic of Communication between VANET and Cloud

The block schematic of communication between VANET and cloud is illustrated in Figure 6.6. Figure 6.6(a) gives information about the data storage model, and Figure 6.6(b) depicts the overall communication process between VANET and cloud. EC order, i.e. compression followed by encryption is applied on VANET data. The VANET data is secured by Standard encryption algorithm. The resultant encrypted data are compressed by a

Table 6.2 Comparison of compression techniques

Compression Technique	CPU Time/ Compression & Decompression Time	Compression Ratio	Throughput	Buffer	Lossless/Lossy	Complexity	Window Size
Brotli	High	High	Moderate	High	Lossless	High	22
Deflate	Moderate	High	High	Moderate	Lossless	Low	15
Zopfli	More	High	Moderate	High	Lossless	Moderate	15
LZMA/LZMA2	More	Moderate	Low	Moderate	Lossless	Moderate	22
LZHMA	More	Moderate	Low	Moderate	Lossless	Moderate	22
BZIP2	More	Low	Low	Low	Lossless	Low	22

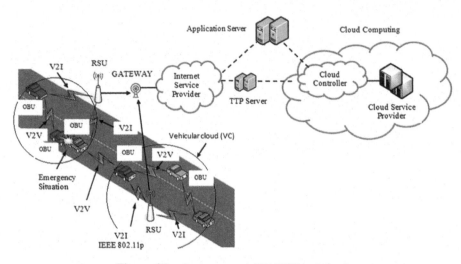

Figure 6.5 Convergence of VANET and cloud.

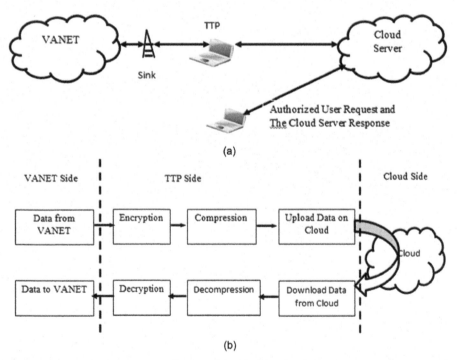

Figure 6.6 Block schematic of communication between VANET and cloud (a) Data Storage Model; (b) System with encryption and compression.

lossless compression algorithm. With encryption and compression system, data owner can outsource his data. Dynamic data will get stored on the cloud server in encrypted and compressed format. File first divided into block and after that each block will encrypt and compress. Data owner can do each block level operation. The authorized user will get the latest version of data when they request for it. Trusted Third Party (TTP) is to create a trust between data owner and Cloud Service Provider (CSP). The detailed explanation about the block schematic is given below.

6.4.3.1 Encryption phase

At the encryption phase, the VANET data are encrypted using Data Encryption Standard Algorithm. The encrypted data can be read only by authorized parties. It protects the confidentiality of messages, and the message content will be denied to the interceptor. Encryption is carried out on the Trusted Third Party (TTP) server. The encryption with compression system is implemented on Amazon Elastic Compute Cloud (Amazon EC2). For Data owner, TTP and cloud server instances of the server are created on Amazon EC2. Windows free tier instances are used for implementation. In an implementation, large Amazon EC2 instance is used to run Cloud server. Amazon cloud instance has configuration as Windows 2008sever, 64-bit base system and Instance ID: windows_server_2008-R2_SPI-English-64bit. One similar type of instance is used for TTP. Data owner and authorized user can login through any device.

6.4.3.2 Compression phase

This phase is used in VANET application to minimize the storage overhead. For compressing VANET data, the Deflate algorithm is used. Deflate is a lossless technique, and It is a combination of two lossless compression techniques namely Huffman encoding and LZ77 algorithm. Compression ratio is more with deflate algorithm. It reduces the storage overhead on a cloud. Compression is carried out on TTP server. Figure 6.7 shows the flowchart of encryption and Compression.

Description of Deflate Algorithm State Transition
The state transition diagram of Deflate Algorithm is shown in Figure 6.8.

1. If a header is available then Deflate algorithm starts in INIT_STATE; otherwise, it starts in BUSY_STATE.
2. From the INIT_STATE, either a dictionary may be set only when we are in INIT_STATE. Then the Deflate algorithm goes in the SETDICT_STATE and one can change the state as mentioned.

3. Irrespective of whether a dictionary is set or not, on start deflates the algorithm goes in to BUSY_STATE.
4. The algorithm is in FINISHING_STATE when flush is called but the process of writing in to the output file is not over. It just indicates the end of the input stream.
5. Deflate algorithm is in FINISHED_STATE when everything is flushed in to the output file.
6. At any other time, the deflate algorithm is in the closed state.

Advantages of Deflate Compression Algorithm

1. Reduces in memory requirement.
2. Reduces in power consumption.
3. Compression is very fast.

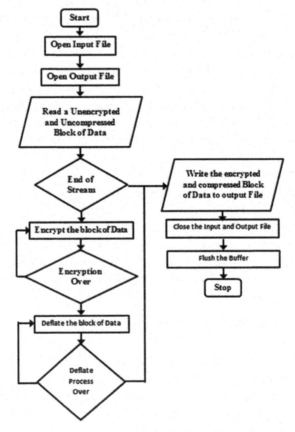

Figure 6.7   Flowchart of encryption and compression of VANET data.

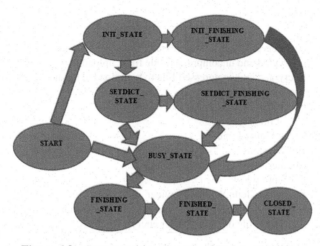

Figure 6.8 State transition diagram of Deflate Algorithm.

6.4.3.3 Cloud
Amazon EC2 cloud instance is used to store the encrypted and compressed data.

6.4.3.4 Decompression phase
This phase is used in VANET application in order to use the compressed file. Inflate is the decompression technique that takes a Deflate bit stream for decompression and correctly produces the original full-size data or file.

6.4.3.5 Decryption phase
In this phase, the uncompressed data that are in encrypted format are converted back to the original form.

The vehicles can make use of V2I infrastructure for accessing external cloud on the move by using the net connection. The vehicles can store important information on the TTP server. Virtualization can be provided by the gateways for all the vehicles to access the TTP server. The data from the TTP server are uploaded on the cloud in the encrypted and compressed format.

6.5 Mathematical Model

Input: {Set of Compressed and Encrypted Files}
Output: {Set of Decompressed and Decrypted Files}
Success: {Successful and secure outsourcing of Dynamic Data}
Failure: {Dispute in integrity of data after outsourcing}

Table 6.3   List of acronyms

Acronym	Meaning
F	Un-encrypted and uncompressed File to be uploaded on the VC
F'	Encrypted and uncompressed File to be uploaded on the VC
ECF	Encrypted and compressed File to be uploaded on the VC
SN	Serial No.
BN	Block No.
n	No. of blocks
BST	Block Status Table
KV	Key Versions
ctr	Counter
EK	Encryption Key
VCSP	Vehicular Cloud Service Provider
$Size_{Ori}$	Size of Original File without encryption and compression
$Size_{Disc}$	Size of the file on the VC after encryption and compression

Table 6.3 gives details about the list of acronyms used.

The un-encrypted and uncompressed File (F) to be uploaded on the VC is divided into n blocks as shown in Equation (6.1).

$$F = \{blk_1, blk_2, blk_3, \ldots, blk_n\} \tag{6.1}$$

The entry of each block is made in BST. BST is a set of tuples that contain SN, BN, and Key Version for each block as shown in Equation (6.2).

$$BST = \{\{SN_1, BN_1, KV_1\}, \{SN_2, BN_2, KV_2\}, \ldots,$$
$$\{SN_n, BN_n, KV_n\}\} \tag{6.2}$$
$$SN_i = BN_i = 1 \text{ and } KV_i = ctr, \text{ for all } i = 1 \text{ to } n$$

As shown in Equation (6.3) and Equation (6.4), F' is the encrypted file generated by encrypting each block.

$$blk'_i = EK(blk_i) \tag{6.3}$$

$$F' = \left\{blk'_1, blk'_2, blk'_3, \ldots, blk'_n\right\} \tag{6.4}$$

Deflate algorithm is used to compress the encrypted file F'. Let us assume ECF as the encrypted compressed File as shown in Equation (6.5).

$$ECF = \text{Deflate\_Compression}\,(F') \tag{6.5}$$

ECF is uploaded on a VC. Now data owner's file F is processed and uploaded on the VC hasctr, KV_i, BST. The VCSP has file, BST.

Now here consider that the size of the original file is $Size_{Ori}$. $Size_{Disc}$ is Actual storage space required.

$$StorageSpace_{Required} = \left(\frac{Size_{Ori} - Size_{Disc}}{Size_{Ori}}\right) \times 100 \tag{6.6}$$

$$StorageSpace'_{Required} = |StorageSpace_{Required}| \tag{6.7}$$

With the help of Equation (6.6) and Equation (6.7), the value of R and $R1$ can be calculated. If the value of R is

$$StorageSpace'_{Required} = |StorageSpace_{Required}|$$

If $StorageSpace_{Required}$ is positive then

$$StorageSpace'_{Required} \propto \frac{1}{storage\ overhead} \tag{6.8}$$

6.6 Simulation and Result Analysis

The proposed system is simulated on Amazon Elastic Compute Cloud (Amazon EC2). Trusted Third Party (TTP) and Cloud Server (CS) instances are created on Amazon EC2. Windows-free tier instances are used for implementation. Large Amazon EC2 instance is used to run Cloud server. Amazon cloud instance has configuration as Windows 2008sever, 64-bit base system & Instance ID: windows_server_2008-R2_SPI-English-64bit-Base_2015.05.1. One similar type of instance is used for TTP. Data owner and authorized user can login through any machine.

A file of size 10 Mb containing the VANET data is encrypted on the TTP server using encryption algorithm. This encryption is necessary for security reason so that only authorized parties can read it. After encryption, the size of the encrypted file has increased by a factor of 0.06%. The encrypted file size is 10.704 Mb. The encrypted file is given as an input to the compression block in order to reduce the storage overhead. After compression, the original file is compressed by a factor of 22%. Figure 6.9 gives a comparison between original file, uncompressed encrypted file, and compressed encrypted file.

6.7 Conclusions

VCC a cloud-based vehicle architecture would increase quality, safety, and security, as well as reduce cost and complexity. The competitive advantage of

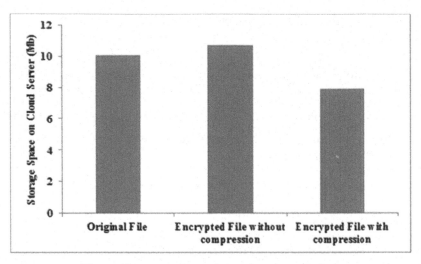

Figure 6.9   Comparison of storage space requirement.

an automotive company would become its ability to enrich the user experience through software innovation. And, most importantly, the VANET would enable the integration of automotive, software, networking, and telecommunications industries, which would lead to an evolutionary transformation of the automotive industry and the world, as we know it. VANET diverges from other types of networks in terms of safety, mobility, network dynamics, a way of communication, etc. Owing to these constraints, utilizing all the VANET resources optimally is a challenging task. A novel idea for the convergence of VANET and cloud is presented in this chapter. The prime goal of the proposal is to reduce storage overhead on the cloud server. The proposed scheme is implemented on the realcloud, and storage cost is calculated. The proposed system has lesser space requirements on the cloud. Taxonomy of VCC is also the significant contribution of this chapter to give scope to VANET and cloud researchers.

References

[1] Al-Sultan, S., Al-Dhoori, M. M., Al-Bayatti, A., and Zedan, H. (2014). A comprehensive survey onvehicular *Ad Hoc* network. *J. Network Comput. Appl.* 37, 380–392.

[2] Toor, Y., and Muhlethaler P. (2008). Vehicle Ad Hoc Networks: Applications and Related Technical Issues. *IEEE Commun. Surveys Tutorials 3rd Quarter* 10, 74–88.

[3] Whaiduzzaman, M., Sookhak, M., Gani, A., and Buyya, R. (2014). A survey on vehicular cloud computing. *J. Network Comput. Appl.* 40, 325–344.

[4] Torrent-Moreno, M., Santi, P., and Hartenstein, H. (2006). "Distributed fair transmit power assignment for vehicular ad hoc networks", in *Proceedings of 3rd Annual IEEE Conference Sensor Mesh Ad Hoc Communication Network (SECON)*, vol. 2, 479–488.

[5] Olariu, S., Hristov, T., and Yan, G. (2013). "The next paradigm shift: from vehicular networks to vehicular clouds," In: *Mobile ad hoc networking: cutting edge directions*, 2nd edn., eds M. C. S. Basagni, S. Giordano, and I. Stojmenovic (Hoboken New Jersey, USA: John Wiley & Sons, Inc.).

[6] Zhang, L., Liu, Y., Wang, Z., Guo, J., and Huo, Y. (2011). "Mobility and QoS oriented 802.11p MAC scheme for vehicle-to-infrastructure communications", in *6th International ICST Conference on Communications and Networking in China (CHINACOM)*, 669–674, doi: 10.1109/ChinaCom.2011.6158239

[7] Sharef, B., Alsaqoura, R. A., and Ismail, M. (2014). Vehicular communication ad hoc routing protocols: a survey. *J. Network Comput. Appl.* 40, 363–396.

[8] Mell, P., and Grance, T. (2011). *The NIST Definition of Cloud Computing.* The National Institute of Standards and Technology, Gaithersburg, MD.

[9] Armbrust, M., Fox, A., Griffith, R., Joseph, A. D., Katz, R., Konwinski, A., Lee, G., Patterson, D., Rabkin, A., Stoica, I., and Zaharia, M. (2010). A view of cloud computing. *Commun. ACM*, 53, 50–58. doi: 10.1145/1721654.1721672

[10] Hussain, R., Son, J., Eun, H., Kim, S., and Oh, H. (2012). "Rethinking vehicular communications: merging VANET with cloud computing," in *IEEE International Conference on Cloud Computing Technology and Science (CloudCom)*, 606–609.

[11] Bouchemal, N., Naja, R., and Tohme, S., (2014). "Traffic modeling and performance evaluation in vehicle to infrastructure 802.11p network", in *Ad Hoc Networks Lecture Notes of the Institute for Computer Sciences, Social Informatics and Telecommunications Engineering* Vol. 129, 82–99.

[12] Andrews, S. (2012). "Vehicle-to-Vehicle (V2V) and Vehicle-to-Infrastructure (V2I) Communications and Cooperative Driving", in *Handbook of Intelligent Vehicles*, 1121–1144.

[13] Harigovindan, V., Babu, A., and Jacob, L. (2012). Ensuring fair access in IEEE 802.11p-based vehicle-to-infrastructure networks. *EURASIP J. Wirel. Commun. Netw.* 168.

[14] Seward, J. "bzip2 and libbzip2, version 1.0.5 A program and library for data compression." Available at: http://www.bzip.org

[15] Andrews, S. (2012). "Vehicle-to-Vehicle (V2V) and Vehicle-to-Infrastructure (V2I) Communications and Cooperative Driving", in *Handbook of Intelligent Vehicles*, 1121–1144.

[16] Harigovindan, V., Babu, A., and Jacob, L. (2012). Ensuring fair access in IEEE 802.11p-based vehicle-to-infrastructure networks. *EURASIP J. Wireless Commun. Netw.* pp. 168.

[17] Bouchemal, N., Naja, R., and Tohme, S. (2014). "Traffic Modeling and Performance Evaluation in Vehicle to Infrastructure 802.11p Network", in *Ad Hoc Networks Lecture Notes of the Institute for Computer Sciences, Social Informatics and Telecommunications Engineering*, Vol. 129, 82–99.

[18] Harigovindan, V., Babu, A., and Jacob, L. (2012). Ensuring fair access in IEEE 802.11p-based vehicle-to-infrastructure networks. *EURASIP J. Wireless Commun. Network.* 168.

[19] Zhang, L., Liu, Y., Wang, Z., Guo, J., and Huo, Y. (2011). "Mobility and QoS oriented 802.11p MAC scheme for vehicle-to-infrastructure communications", in *6th International ICST Conference on Communications and Networking in China (CHINACOM)*, 669–674. doi: 10.1109/ChinaCom.2011.6158239

[20] Armbrust, M., Fox, A., Griffith, R., Joseph, A. D., Katz, R., Konwinski, A., et al. (2010). A view of cloud computing. *Commun. ACM*, 53, 50–58. doi: 10.1145/1721654.1721672

[21] Hussain, R., Son, J., Eun, H., Kim, S., and Oh, H. (2012). "Rethinking Vehicular Communications: Merging VANET with Cloud Computing", in *IEEE International Conference on Cloud Computing Technology and Science (CloudCom)*, 606–609.

[22] Online. Available at: http://google-opensource.blogspot.in/2015/09/introducing-brotli-new-compression.html

[23] Smyrna Grace, O. S., Nalini, T., and Pravin Kumar, A. (2012). Secure and Compressed Secret Writing using DES and DEFLATE algorithm. *IJCSI Intl. J. Comput. Sci.* 9, 423–428.

[24] Alakuijala, J., and Vandevenne, *"Data compression using Zopfli"*. Available at: https://zopfli.googlecode.com/files/Data_compression_using_Zopfli.pdf

[25] Salomon, D. (2006). *Data compression: the complete reference*, 4th edn. Springer, Berlin (ISBN 1-84628-602-5).

[26] Singh, D., Bibhu, V., Anand, A., Maity, K., and Joshi, B. (2014). Study of various data compression tools. *Intl. J. Comput. Sci. Mobile Appl.* 2, 8–15.

[27] Bouchemal, N., Naja, R., and Tohme, S. (2014). "Traffic modeling and performance evaluation in vehicle to infrastructure 802.11p network", in *Ad Hoc Networks Lecture Notes of the Institute for Computer Sciences, Social Informatics and Telecommunications Engineering*, Vol. 129, pp. 82–99.

[28] Miller, J. (2008). "Vehicle-to-Vehicle-to-Infrastructure (V2V2I) intelligent transportation system architecture", in *IEEE Intelligent Vehicles Symposium Eindhoven University of Technology Eindhoven*, The Netherlands, 4–6.

[29] Kakkasageri, M., and Manvi, S. (2014). Information management in vehicular ad hoc networks: A review. *J. Network Comput. Appl.* 39, 334–350.

[30] Dua, A., Kumar, N., and Bawa, S. (2014). A systematic review on routing protocols for Vehicular Adhoc Networks. *J. Vehicular Commun.* 1, 33–52.

[31] Al-Sultan, S., Al-Doori, M., Al-Bayatti, A., and Zedan, H. (2014). A comprehensive survey on vehicular Ad Hoc network. *J. Network Comput. Appl.* 37, 380–392.

[32] Gu, L., Zeng, D., and Guo, S. (2013). "Vehicular cloud computing: A survey", *IEEE Globecom Workshops (GC Wkshps)*, 403–407.

[33] http://fortune.com/2015/09/15/auto-fatalities-increase/

[34] http://www.alertdriving.com/home/fleet-alert-magazine/international/human-error-accounts-90-road-accidents

Biographies

N. Kulkarni received Bachelor of Engineering (B.E.) degree in Electronics Engineering from Walchand College of Engineering, Sangli, Maharashtra (India), in 1996. He has been with Electronica, Pune, from 1996 to 2000. He worked on retrofits, CNC machines as a service engineer, and was also

responsible for PLC programming. In 2000, he received the Diploma in Advanced Computing (C-DAC) degree from MET's IIT, Mumbai. In 2002, he became Microsoft Certified Solution Developer (MCSD). He has over 15 years of experience both in industry and academia. From 2002 onwards, he is working as a faculty in Pune University. From 2002 to 2005, he worked as a faculty in Electronics and Telecommunications Department at GSMCOE, Pune. In 2005, he joined Electronics and Telecommunications Department at RSCOE, Pune, and worked as a faculty till 2007. Since 2007, he collaborates with SKNCOE, Pune, as a faculty in Information Technology Department. He completed Master of Technology (M. Tech) degree in Electronics Engineering from College of Engineering, Pune (India), in 2007. Currently, he is a Ph.D. scholar in Center for TeleInFrastruktur (CTIF), Electronic Systems Department at Aalborg University, Aalborg, Denmark. His area of research is Wireless Sensor Networks (WSN), Vehicular Adhoc Network, and Cloud Computing. He has more than 15 journals and conference publications.

Dr. N. R. Prasad is a security and wireless technology strategist, who through her career has been driving business and technology innovation, from incubation to prototyping to validation. She has focus and the abilities to transform organizations and networking technologies to address changes in markets. She has made her way up the waves of secure communication technology by contributing to the most groundbreaking and commercial inventions. She has general management, leadership, and technology skills, having worked for service providers and technology companies in various key leadership roles.

She is leading a global team of 20+ researchers across multiple technical areas and projects in Japan, India, throughout Europe and the USA. She has been involved in projects and plays a key role from concept to implementation to standardization. Her strong commitment to operational excellence, innovative approach to business and technological problems, and aptitude for partnering cross-functionally across the industry have reshaped and elevated her role as a project coordinator making her the preferred partner in multinational and European Commission project consortium.

Her notable accomplishments include enhancing the technology of multinationals including CISCO, HUAWEI, NIKSUN, Nokia-Siemens, and NICT, defining the reference framework for Future Internet Assembly and being one of the early key contributors to Internet of Things. She is also an expert member of governmental working groups and cross-continental forums.

Previously, she has served as chief system/network architect on large-scale projects from both the network operator and vendor looking across the entire product and solution portfolio covering security, wireless, mobility, Internet of Things, Machine-to-Machine, eHealth, smart cities, and cloud technologies. She was one of the key contributors to the commercialization of WLAN for which she has published two books.

Dr. T. Lin, Distinguished Architect joined Movimento in 2014 after spending a year architecting platforms for a big data platform at the world-renowned Palo Alto Research Center (PARC). From 2012 to 2013, he was vice president of engineering at RingDNA in San Francisco. He also led international development teams as CTO of DigitNexus based in Hong Kong. From 1998 for three years, Dr Lin headed R&D of an award-winning cloud-based enterprise community application platform at Amitive, which was acquired by OpenText. Before that, he spent seven years as a director at SAP Labs in Palo Alto, developing SAP's first industry IoT product: auto-ID infrastructures. Before being an application architect at Silicon Graphics on data mining and visualization. After earning his Ph.D. in computer science from the University of Newcastle in Australia, he was a senior research scientist at CSIRO Mathematical and Information Science in Canberra. Dr Lin earned his M.E. at Beijing University of Science and Technology and his B.E. from Beijing Information Technology University. He holds 15 U.S. patents and has published more than 50 articles and paper.

M. Alam, CTO/CMO Movimento Group Adding deep entrepreneurial skills to international technology management experience and fluency in four languages, Mahbubul Alam is a change driver as CTO/CMO of Movimento. A frequent author, speaker, and multiple patent holder, Alam was brought to the company in early 2015 to reinvent technology and strategy, leading a transformative era in which Movimento will help shepherd the auto industry through the biggest changes since Henry Ford's days. Prior to joining Movimento, Alam spent 14 productive years as a groundbreaking technologist and strategist at Cisco on two continents. He began in the mobile technology arena, later working in a variety of business development and market intelligence capacities in Cisco's Netherlands operation. Having developed a solid name in the company for far-sighted, winning strategies, he was brought to Silicon Valley to head up Cisco's Internet-of-Things (IoT) and Machine-to-Machine (M2M) platforms in 2012.

After coming to California, he proceeded to create a series of now-renowned products and businesses at Cisco: The flagship 4G LTE multi-service router, the IoT edge-cloud gateway, the next-generation Integrated Services Router (ISR), Enterprise Mobility Solutions, and more. Alam grew the company's M2M business from nothing to $350 million in four years. Along the way, he also helped initiate the company's smart connected car roadmap. However, Alam also created several successful internal startups within Cisco, such as IoT vertical solutions for the transportation, manufacturing, and mining industries. He drove corporate investment decisions, ranging between $100 million and $3 billion and directed numerous Cisco portfolios, including the $4 billion Access Routing Technology Group.

Although Alam had created and built a small chain of top-rated dining establishments in the Netherlands that began while he was in college, he moved into the technology arena when he was still in his mid-20s. He worked in research and development for Nokia and later Siemens. There, he led a team that architected the Pan-European ATM (asynchronous transfer mode) network, a Tier-1 ISP that served as Europe's backbone for global Internet access. He also helmed teams that successfully deployed GSM (global systems for mobile communications) for the Dutch railway system as well as launching Europe's first 3G UMTS (universal mobile Telecommunications system) network. Alam was born in Dhaka, Bangladesh, but moved to the Netherlands after receiving a scholarship to attend a prestigious boarding high school. Later, he was awarded degrees in electrical engineering from Holland's renowned Delft University of Technology, with a focus on personal mobile and radar communications.

Professor Dr. R. Prasad is the Founder Director of the Center for TeleIn-Frastruktur (CTIF) at Aalborg University, Denmark, that was established in 2004 and Professor, Wireless Information Multimedia Communication Chair.

Ramjee Prasad is the Founder Chairman of the Global ICT Standardisation Forum for India (GISFI: www.gisfi.org) established in 2009. GISFI has the purpose of increasing the collaboration among European, Indian, Japanese, North-American, and other worldwide standardization activities in the area of Information and Communication Technology (ICT) and related application areas.

He was the Founder Chairman of the HERMES Partnership—a network of leading independent European research centres established in 1997, of which he is now the Honorary Chair.

He is a Fellow of the Institute of Electrical and Electronic Engineers (IEEE), USA, the Institution of Electronics and Telecommunications Engineers (IETE), India, the Institution of Engineering and Technology (IET), UK, Wireless World Research Forum (WWRF) and a member of the Netherlands Electronics and Radio Society (NERG), and the Danish Engineering Society (IDA).

He is a recipient of Ridderkorset af Dannebrogordenen (Knight of the Dannebrog) in 2010 from the Danish Queen for the internationalization of top-class telecommunication research and education.

He has received several international awards such as: IEEE Communications Society Wireless Communications Technical Committee Recognition Award in 2003 for making contribution in the field of "Personal, Wireless and Mobile Systems and Networks", Telenor's Research Award in 2005 for impressive merits, both academic and organizational within the field of wireless and personal communication, 2014 IEEE AESS Outstanding Organizational Leadership Award for "Organizational Leadership in developing and globalizing the CTIF (Center for TeleInFrastruktur) Research Network", and so on.

He is the Founder Editor-in-Chief of the Springer International Journal on Wireless Personal Communications. He is a member of the editorial board of other renowned international journals including those of River Publishers.

Ramjee Prasad is Founder Co-Chair of the Steering committees of many renowned annual international conferences, e.g., Wireless Personal Multimedia Communications Symposium (WPMC), Wireless VITAE, and Global Wireless Summit (GWS).

He has been honored by the University of Rome "Tor Vergata", Italy, as a Distinguished Professor of the Department of Clinical Sciences and Translational Medicine on March 15, 2016.

He has published more than 30 books, 1000 plus journals and conferences publications, and more than 15 patents, and has guided over 100 Ph.D. Graduates and larger number of Masters (over 250). Several of his students are today worldwide telecommunication leaders themselves.

7

Heterodox Networks: An Innovative and Alternate Approach to Future Wireless Communications

Purnima Lala Mehta[1] and Ambuj Kumar[2]

[1]Center for TeleInfrastruktur (CTIF), Aalborg University, Aalborg, Denmark
[2]Global ICT Standardization Forum for India (GISFI), New Delhi, India
E-mail: {pla; in_kumar}@es.aau.dk

Abstract

It is imperative for the service providers to bring innovation in the network design to meet the exponential growth of mobile subscribers for multi-technology future wireless networks. As a matter of research, studies on providing services to moving subscriber groups 'Place Time Capacity (PTC)' have not been considered much in the literature. In this chapter, we present Heterodox networks as an innovative and alternate approach to handle the PTC congestion. In this chapter, we present different approaches to combat the PTC congestion where the traditional terrestrial infrastructure fails to provide sufficient services. Here we define two modes of approach; the first, where the infrastructure is static and resources can be reallocated and the latter is an approach where the infrastructure itself is itinerant. The architecture for the first kind of deployment is termed as 'Self-Configurable Intelligent Distributed Antenna System' that overlays intelligence over the conventional DAS architecture and the latter is in the form of a swarm of intelligent hovering base stations to relieve the moving user congestion. A suitable network architecture of a 'Hovering Ad-hoc Network' for the latter will be deployed to assist and manage the overloaded primary base stations enhancing the on-demand coverage and capacity of the entire system. Proposed modes can either operate independently or as a cascaded architecture to form a Heterodox Network.

Role of ICT for Multi-Disciplinary Applications in 2030, 149–166.

Keywords: Place time capacity, Heterodox networks, SCIDAS, HANET.

7.1 Introduction

Current reports suggests more than half a billion (563 million) mobile devices and connections were added in 2015 with smartphones accounting for most of that growth. Global mobile data traffic reached 3.7 exabytes per month at the end of 2015, up from 2.1 exabytes per month at the end of 2014 [1]. The LTE small-cell architecture brings upon big challenges with itself in effectively organizing, managing and optimizing the network resources to the emerging Heterogeneous Networks or so-called HETNETs [2]. Formidable user traffic coverage and capacity requirements are becoming high-reaching challenges and to service this humungous user-traffic data, the network operators need to unlock ingenious paradigms into future generation networks.

Amidst these challenges, serving data hungry outdoor user hotspots is a significant problem to be addressed by the network operators. The challenges in relieving the congestion, offloading the nearby base station with offering the best data rates to the user demand with minimal dropped calls is a concern. These problems become even more challenging to serve when the user hotspots are mobile. These moving user hotspots create demand in coverage and capacity at every position they traverse and has been coined in as the problem of 'place time capacity (PTC)' [3]. Moving hotspots expect high signal quality and data rates whenever and wherever they are. Such situations are observed frequently in highly populated countries like India where subscriber grouping is common and slight variation in routine day-to-day characteristics leads to huge accumulation of potential subscribers at single or multiple locations. Deploying temporary solutions like network on wheels (coverage on wheels, etc.) or pico cell deployments have been adopted in past. However, being static/quasi-static in nature, such solutions are not sufficient for the subscribers that are predominantly dynamic in nature.

In this chapter, we present an insight to 'Heterodox networks', a shift in paradigm to conventionally deployed solutions to handle PTC situations. We firstly define the concept of PTC and then present two possible solutions to resolve it, namely, (1) Self-Configurable Intelligent Distributed Antenna System (SCIDAS) and (2) Hovering Ad-hoc Network (HANET). We describe the shortcomings of the present hardware systems in each section before describing the concepts. Lastly, we describe a holistic approach of SCIDAS–HANET as a cascaded solution to solve PTC congestion.

7.2 Concept of PTC

The accumulation of subscribers and their movements together in groups/clusters create an apparent capacity demand at every location they visit. We have defined this challenge in a novel way as PTC'. Although, the cellular networks are designed to cater the subscriber's mobility, however, subscriber accumulation due to certain motives or triggers might increase the capacity demand beyond the servicing limit of a base station. The subscribers having common interests, accumulate, and move in groups, thereby increasing the probability of finding high capacity demand at different places during the event. We can observe this phenomenon in carnivals, marathons, festivals, etc., where using cell phones and other mobile terminals is vital to the subscribers. We see PTC problem as one of the major concerns that needs to be considered for the future generation networks and in the following sections we present the possible solutions to combat it.

7.3 Self-Configurable Intelligent Distributed Antenna System

7.3.1 Background

It is evident that with the forthcoming telecom generations, the present cellular technologies will go hand in hand due to slow upgrade of the newer network by the service providers. It takes years for a new technology to be absorbed by the user base completely and yet by that time a newer technology seeps in. Talking of the state-of-the-art LTE technology, with the increasing number of subscribers, technologies like Distributed Antenna System (DAS) and Cloud-Radio Access Network (C-RAN) are presently operative to enhance the network area coverage and capacity [4–6].

On the other side, we see there are certain shortcomings with these technologies:

- Infrastructure dependent and hence difficulty in adaptation with newer future technologies.
- Limited scope of C-RAN in dense-nets in laying high-capacity fibers at every point in the service area called as Area-of-Interest (AoI) in the SCIDAS concept.
- C-RAN is a base station centric solution where subscribers are tied to a particular base station. This means no mobility of subscribers are taken into concern.

- Dynamic group user behavior in terms of mobility (PTC) is not considered in these technologies.
- Remote units do not work for physical restructuring of the network for the betterment of the network environment and subscriber needs in the AoI.

To overcome the above mentioned challenges and to accommodate the place time capacity problems, Ambuj et al introduced a concept of 'SCIDAS' as an intelligent architecture over the traditional DAS architecture [7].

7.3.2 SCIDAS Architectural Components

The significance of SCIDAS lies with the ability to respond to dynamic network conditions at each and every remote unit distinctively, exclusively and simultaneously. The different components of SCIDAS architecture are:

- SCIDAS Node (SCIN): BTS hotel equivalent in SCIDAS responsible for all the boss functions like planning, gathering, deciding, and configuring the global network functions.
- Network Intelligence Unit (NIU): An entity residing in SCIN that adds technology-independent intelligence functionality layer to the present DAS paradigm. Two types of NIUs, Boss (NIUB) and Sub (NIUS).
- Follower Unit (FU): It is the equivalent of remote radio head (RRH) but with multi-technological support. The radio part resides only in FU enabling NIU to select the individual carriers/subcarriers and to feed an individual antenna differently. FU takes instructions from NIU.
- QUICKNET Access Port (QAP): It is an easy deployment option for inaccessible areas (in terms of laying optical fiber) and/or sporadic network scenarios as a sub-network system (branched expansion) that is called as QUICKNET. A SCIN with its respective SCIWAN is termed as SCICELL. This SCICELL is a building-block that will repeat in the entire AoI.
- Two distribution network layers, (1) Neuron Network (NuN) that connects the FUs through fiber optic network/wireless access ports and, (2) Spine Network (SpiN), connects the SCICELLs to each other through a high capacity network.
- Maneuvrable and Controllable Platform (MCP): The docking station for the antennas that resides at the FU and handles the physical conditioning of the network by orienting and tilting the antenna of every sector of the remote unit. MCP is controlled by the NIU via the FU by sending instructions to the FU.

- Global User Bit Handler (GUBH): A type of base station with RF and modulation capabilities and manages user data bits.
- Active Probing Management System (APMS): Allows a network to have a panoramic view of the network and predict the future subscriber behaviour.

Figure 7.1 shows the architectural components of SCIDAS.

7.3.3 Understanding SCIDAS

The SCIDAS network follows the C-RAN architecture that is eventually an evolution of a traditional DAS to a higher level. Adding intelligence to the traditional DAS yields SCIDAS, wherein the intelligence implicates: (i) the capability to understand the varying physical conditions that impact the communication network environment in both optimistic and pessimistic ways and to identify the epicenter of problems in case network conditions deteriorate and, (ii) the capability to plan and decide the changes, whether mechanical or electrical, which should be implemented across the network to cope up with the deteriorating conditions so that the network performs at or above a certain health benchmark in every condition and at every time.

SCIDAS contains a node that contains the necessary equipment of the network and also represents the Network Core. There are two kinds of nodes in the SCIDAS architecture namely the Boss and Subordinate and the name defines the hierarchical dependency of nodes on each other. There can be multiple Subordinate nodes but only one Boss node to control everything. A Boss can share or transfer its properties to a Subordinate node depending upon the present network scenario conditions; however, the key controls are retained by it. Nonetheless, unlike the usual Base Station Pool (BSP), a SCIDAS node has GUBH which is a technology independent base station, from user's perspective, and is involved in managing the user bits irrespective of the technology they use. As seen in Figure 3-1, the GUBH resides in both Boss and Subordinate nodes and its role is very much similar in both the cases. Hierarchically, GUBH residing at the Boss is the superset of all the GUBHs at various Subordinates. This means that properties of all the GUBHs at subordinate level are combined and stacked at the GUBH at Boss in addition to the properties owned by the GUBH. A GUBH holds the channel information and user bits associated with it.

Similar to the present C-RAN/DAS architecture, all the sites in SCIDAS have a FU, similar to a Radio Remote Unit (RRU), prefixed to the antenna system that are responsible for following the NIU and to configure the

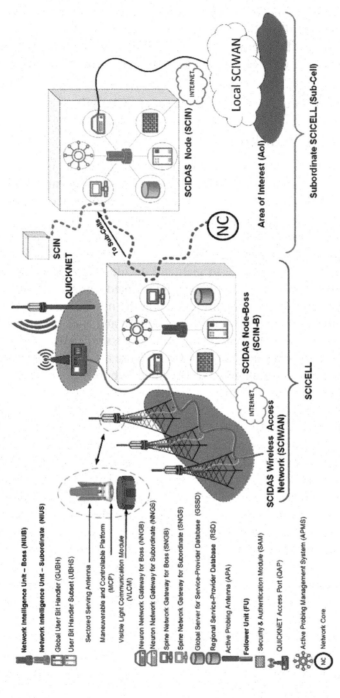

Figure 7.1   SCIDAS architecture [7].

behavior, and to control the performance of sites (such as resource allocation and carrier aggregation) according to the network dynamics at the very location. In the SCIDAS, the intention is to separate the user data from the technology; therefore, all the channel modulation and link related techniques are performed at FU end. This makes SCIDAS more adaptive to the network dynamics and the FU being a controllable module (controlled by NIU), can be instructed to behave independently form others thereby enhancing the channel utilization and improvising Quality of Service.

The communication of NIU with FU is done in two stages, block level and local level. An NIU manages the GUBH resources and spreads across the SCIDAS network via optical fiber network (OFN) (similar to C-RAN/DAS) that we termed in Kumar et al. [7] as NuN (shown as bold purple curve in the Figure 7.1) through Neuron Network Gateways (NNGs). An NNG is similar to master unit (MU) in conventional DAS that does the necessary conversion of data into optical signal to be spread across SCIDAS through NuN; however, unlike the conventional MU, NNG also assigns the address of the corresponding FU to the packets that are thrown to the NuN. This makes packets to arrive only at the destined FUs benefiting the spectrum management. FUs, the Node, and NuN network altogether is a Block and the access network part of block is SCIDAS Wireless Access Network, or SCIWAN. The network deployment must begin first with Boss Block that includes Boss node and it is SCIWAN that is termed in the discussed work as SCIWAN-B. Growing SCIWAN beyond a considerable limit may lead to significant delays in the packets arrival of far flung remote units. To avoid this situation, SCIDAS has a provision to "Copy and Paste" the node at another strategic location and let the rest of SCIWAN grown under its supervision. The new node shall, however, be lower in the hierarchy to the Boss and, as discussed earlier, is a Subordinate node. The Boss is connected to its subordinates through another high capacity network termed here as Spine (shown as dotted red curve in Figure 7.1). Just like NNG, the Spine also has gateways, called as Spine Network Gateway or SGN, however, located at either nodes differentiated as SGN for the Boss or SGNB and SGN for a Subordinate or SGNB.

7.3.4 SCIDAS in Handing PTC

As explained in the previous section, PTC refers to the dynamic behavior of the users that create capacity wobbles at every point of traversal. As the base stations are designed to handle a certain capacity at any given time, the contingency of PTC generates network congestion resulting in overloading the nearby base stations. The SCIDAS initially segregates the user traffic on the basis of multiple-technologies used by them. The FU in each technological layer will be allocated with maximum number of carriers to serve the generated traffic. The adjacent FUs can be declined of certain carrier allocation to avoid interference still maintaining the subscriber connectivity. The active probing determines the user movement behavior and the APMS will predict the next affected area. Once the locations of accumulations are determined, the FUs in that area will automatically configure themselves to serve the incoming PTC traffic and this goes on in an iterative manner. Hence, the intelligent and proactive components in SCIDAS manage the PTC bursts within the interference limits.

With SCIDAS, capacity bursts due to accumulated subscriber clusters floats above in a higher layer from beginning until it is dispersed without bothering the underneath network deployment. The SCIDAS architecture allows NIU to allocate the carriers exclusively and variably for every FU and, therefore, making the dynamic allocation of carriers to relevant FUs feasible. Hence, sites that are suffering from PTC congestion can be allocated with more carriers maintaining the healthy signal to interference noise ratio by lowering the carriers in the adjacent lowly utilized sites carriers making channel utilization more efficient. This resource allocation can be conveniently changed as the cluster moves, divides or disperses during course of time.

7.4 Hovering *ad hoc* Network (HANET)

7.4.1 Background

Ongoing industrial projects like Facebook Drone [8] and Google Loon [9] are in progress in laying high-altitude platforms to extend web access and internet connectivity to the ground users. On the other side, small unmanned aerial vehicles are being considered to be the most interesting and promising solution to both civilian and military applications. Low-altitude aerial platforms are potentially believed to be deployed as alternatives to fixed wireless infrastructure in providing cellular services to the ground users especially for public safety communications during disaster recovery and emergency services. Recent projects like AVIGLE [10, 11], SMAVNET [12], ABSOLUTE [13–15], etc., are also dealing with UAV/Drone based implementation in the areas of surveillance, disaster management, public safety (emergency services), and cellular communication services. Small aerial platforms have also been proposed to improve the coverage and capacity in the state-of-the-art cellular networks [11, 16].

Talking about hotspot conditions, a few authors have considered moving subscriber hotspots in literature. In Claussen [17], definite number of adaptive base stations were set up on a rail at the ceiling of an airport building to provide cellular service to the moving crowd at the arrival of an airplane. A cell outage condition was outlined in Rohde and Wietfeld [11] wherein a swarm of UAVs equipped with LTE-4G were utilized to offload traffic into neighboring cells and further suggested that small UAVs can be used for moving traffic overload situations (parade, etc.,). A statistical propagation model for predicting the air-to-ground path loss between low-altitude platforms (LAPs) and terrestrial terminal was proposed in Al-Hourani et al. [14] to enhance studies related to user traffic offloading during massive moving of crowds. The base stations are designed to service a certain amount of traffic at any given time. Sudden high traffic bursts due to PTC can make it difficult for the base stations to deliver the required services with high data rate and full connectivity. We see that no specific case in managing cellular services by aerial base stations particularly to moving crowd situations has been addressed in any of the papers.

Earlier Purnima et al. [18] proposed a concept of HANET as an *ad hoc* network of airborne intelligent base stations to serve the PTC congestion situations. A HANET was coined as a team of self-itinerant, intelligent and, adaptive hovering radios to serve the dynamic traffic need of moving subscriber hotspots by optimally positioning themselves at the Area of Event (AoE) where the moving crowd accumulation is potentially high. These HANET members will follow the ground subscriber groups with the aim of covering the AoE where the conventional approach is lagging or failing to serve and, leaving no possibility of coverage voids and overlaps. This concept proposes HANET to work as an assistive network to the primary network to offload the overloaded ground base stations with the aim of relieving the PTC congestion.

The HANET is a specialized form of *ad hoc* network that mainly differs from previously defined VANET and FANET [19] in having lower mobility, LOS connectivity between the devices in the HANET, close to the ground and ability to provide cellular services to the ground mobile users.

7.4.2 Why to Deploy HANET for PTC?

In the following, we give some points that supports our ideology in exploiting airborne base stations in serving the PTC groups,

- Low-altitude aerial platforms can be rapidly deployed at the AoE.

- Dominant LOS path between HANET member and ground users eliminating most of multipath and shadowing effects.
- Team of such aerial platforms can be established on ad-hoc basis forming a wireless mesh network across the AoE.
- They can act as relays to offload the user traffic and also to extend the coverage of ground base stations.
- With HANET members following the PTC groups, the dropped calls due to multiple handovers will reduce.
- These aerial devices are a solution to low cost deployment saving the Capital Expenditure (CAPEX) and Operational Expenditure (OPEX) in installing, constructing and maintaining the conventional cellular towers.
- These devices can also be incorporated with intelligence, and can be customized as per the need of the network.
- Small size, portability and advanced technology makes it an apt choice to be deployed easily.

7.4.3 HANET Network Architecture

There are some additional network elements that we have added to modify a conventional overloaded hotspot scenario as shown in Figure 7.2. As can be seen in the figure, there are two base station BS1 and BS2 that are overloaded beyond their capacity deliverance due to PTC gathering. This entire region comprising of PTC groups and nearby base stations is the AoE. First, we define the HANET members that are quad-copter based aerial drone devices, equipped with BTS facility and antennas to be able to provide cellular services to the ground users. The second network element we describe here is the HANET Gateway Base Station (HGBS) connected directly to the backhaul.

The HANET members will be operating as HANET Serving Member (HSM) or HANET Relay Member (HRM) or both at any instant of time depending on their positions across the AoE. The HSMs will be responsible to provide cellular services to the ground users and the HRMs will be taking care of the user data that will be eventually relayed to the HGBS to be connected to the backbone network. Hereafter, we define the HANET Base Station Subsystem (HBS) comprising of the HSMs and the overloaded base stations and HANET Relay Subsystem comprising of the HRMs and the HGBS. The HGBS is a base station close to the AoE, not to be utilized to provide the cellular services to the PTC but connected with the HANET relay network to send the user traffic to the backbone network.

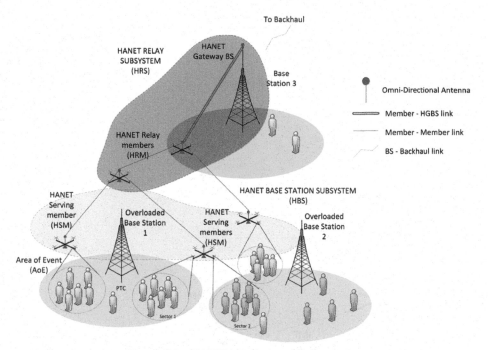

Figure 7.2 HANET network architecture [18].

For the team of HANET members, it is required to cooperatively provide cellular coverage to the ground users; therefore, it is essential for the team members to communicate. There will be typically three types of communication existing for controlling and managing the entire network,

- *Member to User Communication* – The HANET is employed to service the ground users beneath and therefore requires communication channels to provide services as a part of the data plane. We define two communication channels namely the IMT link for state-of-the art cellular services and a millimeter wave (mmW) link for the future phones expected to be working in the millimetre frequency range.
- *Member to Member Communication* – For the network management it is crucial for the members to communicate among themselves to be able to take decisions on a team basis. The control information like location of each member, member IDs, etc., for deciding on optimal positioning and maneuvring control is proposed to be communicated through a separate communication layer based on the MMW channels so that there is no interference with the IMT user data.

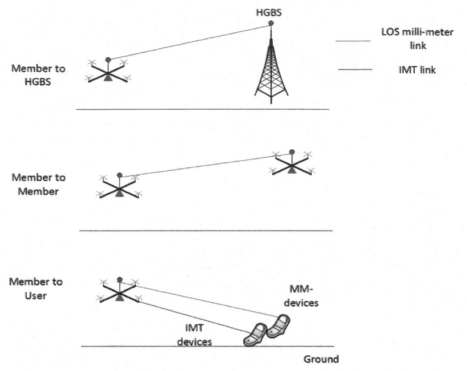

Figure 7.3   HANET communication links [18].

- *Member to Base Station Communication* – The HANET members need to communicate with the existing primary network and the HGBS to relay the user data and also to share the network information as a part of the control plane. This part of communication is also proposed to be shared through MMW links.

Figure 7.3 illustrates different communication links between the network elements in the HANET network architecture.

7.5 Heterodox Network: Cascaded SCIDAS–HANET Architecture

The SCIDAS network discussed in the Section 7.3 is a terrestrial deployment and, although it has high intelligence-based dynamic and strategic resource allocations, its approachability is limited by its fixed infrastructure. Moreover, SCIDAS allocates the carriers that are pooled in GUBH that are allocated to service providers by regulators. A GUBH is a global base station that receives the user-data bits irrespective of the technology. GUBH, assigns address of the target FU to the packets that belongs to user underneath the FU. This enables SCIDAS to choose only those FUs that are catering intended subscribers. Therefore, the resource allocation can be made individually and differently for every FU in the SCIDAS service area. Any excessive influx of people (such as gathering of people from different countries during a carnival), from areas other than those belonging to the AoI, may cause deficiency in allocating the carriers from the pool as total capacity demand of the entire network may surpass the capacity offered by the network and, the SCIDAS might underperform in such cases. Also, there may be situations when the SCIDAS is unable to cater users; such situations arise when users are either beyond coverage footprints of the SCIDAS network, or residing in the coverage hole within the serving network. To mitigate both capacity and coverage constraints of SCIDAS, the rapid deployment architecture, HANET, which is a self-itinerant intelligent radio architecture (SIIRA), is amalgamated with the SCIDAS architecture.

Figure 7.4 shows a heterodox network that cascades the two aforementioned architectures that are discussed in Sections 7.3 and 7.4 operating in a complementary mode. This enhances the network deployment to a considerably rapid level. The HANET system can be rapidly deployed at AoE with stringent situations where they create sub-cells within SCICELL to offload cater the tedious situation and backhaul to the nearest HGBS.

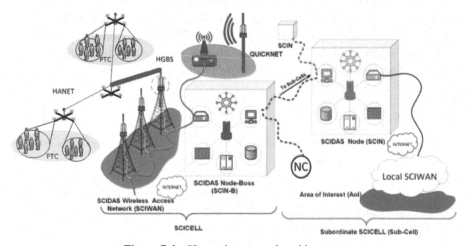

Figure 7.4 Heterodox network architecture.

In Heterodox network, APMS can work with HANET system to monitor and follow the PTC more conveniently than a conventional terrestrial network. As SCIDAS has capability to conveniently allocate carriers to an individual site exclusively, the site that is performing as HGBS can be dynamically allocated with more lines for rapid backhauling that can itinerate with the movements of HANET members.

This composite architecture may prove to be quite resilient to various adversities. For instance, HANET member may take charge as a temporal primary network in case the base SCIDAS network fails to operate due to natural disasters (flood, earthquake, etc.), sabotage, or, technical faults. The SCIDAS network can approach non-deployable areas through its wireless backhauled sites called here as QUICKNET in which, the FU NIU are connected wirelessly though QAP. This QAP may also facilitate backhauling of a HANET relay in absorbing the capacity burst in the concerned area.

7.6 Conclusion

The exploding user traffic demand is increasing exponentially and posing challenges to the service providers especially for the moving hotspot conditions which are the PT groups. We have presented Heterodox network, as a combination of two solutions to combat PTC namely: (1) SCIDAS having a fixed infrastructure and, (2) HANET with dynamic infrastructure solutions. We have discussed each of the two kinds as extension of previous works and how they can be combined together to have a holistic solution. We have also discussed how two solutions complement each other; SCIDAS following terrestrial deployment strategy, while HANET complimenting the limitations of SCIDAS for its fixed infrastructures. The concept discussed here are endorsed diagrammatically in this chapter.

References

[1] CISCO. (2016). *Cisco Visual Networking Index: Global Mobile Data Traffic Forecast Update*, 2015–2020 White Paper.
[2] Wang, X., Li, X., and Leung, V. C. M. (2015). Artificial intelligence-based techniques for emerging heterogeneous network: state of the arts, opportunities, and challenges. *IEEE Access* 3, 1379–1391. doi: 10.1109/ACCESS.2015.2467174.

[3] Kumar, A., Mehta, P. L., Prasad, R. (2014). "Place time capacity: a novel concept for defining challenges in 5G networks and beyond in India," in *2014 IEEE Global Conference on Wireless Computing and Networking (GCWCN)*, 278–282.

[4] Huang, J., Duan, R., Cui, C., and Chih-Lin, I. (2014). "Overview of cloud RAN," in *2014 XXXIth URSI General Assembly and Scientific Symposium (URSI GASS)*, 1–4.

[5] Beyene, Y. D., Jantti, R., and Ruttik, K. (2014). Cloud-RAN architecture for indoor DAS. *IEEE Access* 2, 1205–1212.

[6] Niu, H., Li, C., Papathanassiou, A., and Wu, G. (2014). "RAN architecture options and performance for 5G network evolution," in *2014 IEEE Wireless Communications and Networking Conference Workshops (WCNCW)*, 294–298.

[7] Kumar, A. Mihovska, A., and Prasad, R. (2015). "Self-configurable intelligent distributed antenna system for resource management in multilayered dense-nets", in *Accepted in Wireless VITAE, 2015 Global Wireless Summit* (In Print).

[8] Facebook. (2014). *Connecting the World from the Sky*. Technical Report.

[9] Project Loon. (2015). *Wikipedia, the free encyclopedia.*

[10] Rohde, S., Goddemeier, N., Wiefeld, C., Steinicke, F., Hinrichs, K., and Ostermann, T., et al. (2010). "AVIGLE: a system of systems concept for an avionic digital service platform based on micro unmanned aerial vehicles," in Proceedings of IEEE International Conference on Systems, Man, and Cybernetics (SMC).

[11] Rohde, S., and Wietfeld, C. (2012). Interference aware positioning of aerial relays for cell overload and outage compensation," in *2012 IEEE Vehicular Technology Conference (VTC Fall)*, 1–5.

[12] S. Hauert, S. Leven, J.-C. Zufferey and D, Floreano. (2010). "Communication-based swarming for flying robots," in *International Workshop on Self-Organized System*, Zurich, Switzerland,.

[13] "EU-FP7 ICT IP Project ABSOLUTE." (2013). Available at: http://www.absolute-project.eu/reports/publications

[14] Al-Hourani, A., Kandeepan, S., and Jamalipour, A. (2014). "Modeling air-to-ground path loss for low altitude platforms in urban environments," in *2014 IEEE Global Communications Conference (GLOBECOM)*, 2898–2904.

[15] Al-Hourani, A., Kandeepan, S., Lardner, S. (2014). Optimal LAP altitude for maximum coverage. IEEE Wirel Commun. Lett. 3, 569–572.

[16] Guo, W., Devine, C., Wang, S. (2014). "Performance analysis of micro unmanned airborne communication relays for cellular networks," in *2014 9th International Symposium on Communication Systems, Networks & Digital Signal Processing (CSNDSP)*, 658–663.

[17] Claussen, H. (2005). "Autonomous Self-Deployment of Wireless Access Networks in an ionAirport Environment," in the Proceeding of 2nd Internat. Federation for Informat Processing (IFIP) Interntional Workshop on Autonomic Communication.

[18] Mehta, P. L., Sorensen, T. B., and Prasad, R. (2015) "HANET: millimeter wave based intelligent radio architecture for serving place time capacity issue," in *Wireless VITAE, 2015 Global Wireless Summit* (In Print)

[19] Bekmezci, İ., Koray Sahingoz, O., Temel, Ş. (2013). Flying *ad-hoc* networks (FANETs): a survey. *Ad Hoc Netw*. 11, 1254–1270. doi.org/10.1016/j.adhoc.2012.12.004.

Biographies

Assistant Professor P. L. Mehta has received her MTech at NorthCap University (formerly ITM University), Gurgaon, India. Currently she is Assistant Professor at Electronics and Communication Department at HMR Institute of Technology and Management, Delhi, India. She is presently pursuing her Ph.D. under the Centre for TeleInfrastruktur (CTIF) section, Department of Electronic Systems, Aalborg University (Denmark) as a part of the GISFI Ph.D. scholarship program. The major fields of scientific interest include *ad hoc* networks, aerial base stations, millimeter wave-based devices.

Research Assistant A. Kumar has received Bachelor of Engineering in Electronics and Communications from Birla Institute of Technology (BIT), Ranchi, India in the year 2000. He worked in several leading multinational telecommunications companies including telecom service providers until 2009. Thereafter, during 2009, he worked as Research Associate at the Centre for TeleInfrastruktur (CTIF) section, Department of Electronic Systems, Aalborg University (Denmark). He has received research scholarship under the European Commission-Erasmus Mundus "Mobility for Life" and currently perusing his Ph.D. work since 2010 at the CTIF section, Aalborg University (Denmark). His research interests are radio wave propagation, cognitive radio, visible light communications, and distribute antenna systems etc. He has several research publications in these areas.

8

Network Neutrality for CONASENSE Innovation Era

Yapeng Wang and Ramjee Prasad

Center for TeleInFrastruktur (CTIF), Aalborg University,
Aalborg, Denmark

Abstract

In 2015, Federal Communication Committee (FCC) and European Commission enacted respective rules in relation to open Internet, the network neutrality (as commonly known) or Net Neutrality (as in Europe), which gained more attention, and the topic was discussed ardently. This paper reviews the development of NN debate process, and the opinions from different sides, including the network providers, the service providers, other relevant companies, governments, and researchers. This paper also introduces a telecommunication convergence concept, that is Communication, Navigation, Sensing, and Services (CONASENSE). It aims to formulate a vision on solving societal problems with a new telecom technique to improve human welfare benefit. This paper focuses on the service of CONASENSE and summarizes the current situation of NN in service innovation era.

Keywords: Network Neutrality, CONASENSE, Innovation, Long term benefits.

8.1 Introduction

The Information and Communication Technology (ICT) is playing an important role in our life. It supplied us with many various innovative services and applications, from e-commerce and e-health to real-time telephone meetings with live video streaming, improving human's Quality of Life (QoL) and benefiting the whole society. We have entered a service innovation era, in which every part of the industry chain is making contribution to the ICT industry. Internet Services Providers (ISPs) are continuing to upgrade the network, the Internet Content Providers (ICPs) are supplying wide range of various content and services to keep the telecom industry prosperous. How these innovative services run over the telecommunication network is governed by not only technology, but also by the rules as proved by government or authorities in some region. NN, as an important regulation which influences future Internet development, aims that every end user has the equal right to access the Internet and use the legal Internet content and applications. CONASENSE, as will be discussed later in this chapter, is a telecom convergence concept that will be run above the ICT platform. This chapter focuses on the CONASENSE service, and analyzes NN rules' impact on this service.

8.1.1 Network Neutrality

Network Neutrality rules aims to provide an open Internet [1] to the end users. The open Internet (by FCC defined) refers to *"uninhibited access to legal online content without broadband Internet access providers being allowed to block, impair, or establish fast/slow lanes to lawful content"* [2].

This means that legal content, whether it is an application or data, must reach users without the intermediate communication system controlling its flow. Internet Content Providers (ICPs) and Internet Service Providers (ISPs) cannot block, throttle or create the special facilities for a content or application. This gives a kind of liberty to the end users to enjoy the variety of information without bothering about how ICT is dealing with that information. Therefore, users may demand a better network services to enjoy the lawful contents. Section 8.2 introduces NN in detail.

8.1.2 CONASENSE [3]

CONASENSE refers to Communication, Navigation, Sensing and Services. In November 2012, CONASENSE foundation was established to support its development, as an integration of communications, navigation and sensing technology. It helps define and steer processes directed towards actions on investigations, developments and demonstrations of service innovation, especially for those services that have high potential and importance for society. Section 8.3 gives a detailed introduction to this concept.

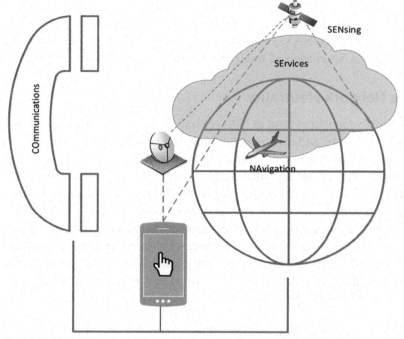

Figure 8.1 CONASENSE framework [3].

8.1.3 Innovative Services

In this chapter, the innovation service refers to over the top service (OTT). OTT implies communications carried over the physical network infrastructure using an IP protocol to reach services available on the Internet [4]. An OTT application is any app or service that provides a product over the Internet and bypasses traditional distribution. Services that come over the top are most typically related to media and communication and are generally cheaper than the traditional method of delivery. As we discussed CONASENSE service belongs to the OTT service domain. Section 8.4 will discuss NN's impact on it.

This chapter points out that both the opponent and proponent of NN debate agree on the need to keep the Internet open to innovation, and preserve the freedom of users to access the content and services. Meanwhile, this chapter also analyzes how NN impacts innovation in CONASENSE, meeting the regulation goal. This chapter makes analysis mainly from technical perspective; additionally, social and economic aspects are also discussed.

Apart from Section 8.1, the rest of the chapter is organized as follows: Section 8.2 reviews the evolution and current discussions of NN, Section 8.3 briefly introduces CONASENSE, Section 8.4 qualitatively analyzes the impact of NN on Services, and finally Section 8.5 concludes the chapter.

8.2 The Network Neutrality

More and more regulators tend to believe that the ICPs are the motivation of innovation, economy, and investment. Combined with the rapid development of the Internet as a ubiquitously available platform and resource, the network infrastructure owners are regarded as to *"have both the incentive and the ability to act as gatekeepers standing between edge providers and consumers. As gatekeepers, they can block access; target competitors, extract unfair tolls"* [2]. But the open Internet is regarded as to be the guarantee to the innovation, economy, and investment; therefore, more and more countries government enacted the NN rules. Presently, more than 10 regions enacted relevant rules [5]. Figure 8.2 shows the regions that either have enacted NN or are in their discussion phase.

8.2.1 The Concept of NN

The concept of NN indicates that Internet service providers make or keep the Internet open, and ensure all the users have same right to access the network and use the content or services without any discrimination [2].

8.2.2 NN's Principles

Currently, Network Neutrality is a global debate [5]. FCC released its updated Open Internet Order in 2015, to enact strong, sustainable rules to protect the open Internet. The order includes 3 bright-line rules as follows [2]:

- No Blocking (NB), to prohibit the network providers to block the legal applications, contents and devises.
- No Throttling (NT), to prohibit the network providers to degrade the traffic of legal applications and contents.
- No Paid Prioritization (NPP), to prohibit the network providers to provide and charge the differentiated service for the applications and contents.

"No unreasonable interference or disadvantage to consumers or edge providers [2], *and, enhanced transparency between users and ISPs". "As with the 2010 rules, this Order contains an exception for reasonable network management, which applies to all but the paid prioritization rule"* [2] is also emphasized in the FCC rules.

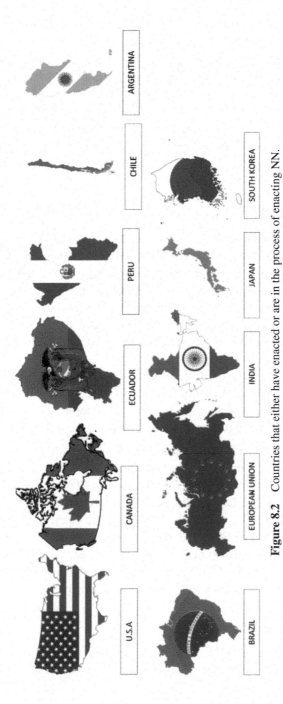

Figure 8.2 Countries that either have enacted or are in the process of enacting NN.

In October 2015, the European Parliament also approved the first EU-Wide NN rules that enshrined its principle into EU law [6]. It indicates that: *"No blocking or throttling of online content, applications and services are allowed. Accordingly, every European must be able to have an access to the internet and, all contents and services, via a high-quality service that is provided by an ISP such that all traffic must be treated equally. NN rules seems more inclined to the end users and, equal treatment allows reasonable day-to-day traffic management according to justified technical requirements which must be independent of, the origin or destination of the traffic, and, of any commercial considerations."*

8.2.3 History of NN

The phrase of Network Neutrality was first proposed in a law review article [7] by Tim WU in 2003, mainly referred to the concepts of freedom, competition, and innovation. NN suggests that each network protocol layer should be independent and perform the assigned duties at the original phase. With the expansion of the commercial Internet, the focus of the market competition has been shifting from the connection and network layer to the application and content layer. The main debate of NN also was switched from technological field to commercial field.

Through the key events (as follows) that happened in the USA, we can find that the way of NN is not smooth. Meanwhile, the term holds different meanings to different expertise, and NN has different debate focuses in different phases. The discussion of NN will be introduced later on.

- In December 2010, The FCC approved the Open Internet Order which consisted of three items of NN regulations; they are *"Transparency, No Blocking and No unreasonable discrimination"* [8].
- In January 2014, the United States Court of Appeals for the District of Columbia Circuit (D.C. Circuit) overturned the Open Internet Order.
- In February 2015, The FCC issued Open Internet Rules and Order, and in June 2015, The Open Internet Rules and Order came into effect officially.
- In August 2015, The D.C Circuit announced the Open Internet Rules and Order and will face an important federal-court test.

8.2.4 Current Discussion on NN

From the history of Network Neutrality, we can find that NN has been really controversial. In 2015, European Commission and FCC enacted the NN rules.

Most of the goal and the principle in NN rules are very concurrent, but the distinction still existed. The essential debate between Europe and the USA will be discussed below:

8.2.4.1 Service innovation

The proponents of NN are mainly those enterprises that are related to Internet contents. Worriedly, they are stating that actions departing from NN principles could threaten the innovation of the Internet content as, ISPs may increase control on the content and applications over the Internet. In order to encourage the innovative services with enhanced Quality of Service (QoS) especially from startups, the new EU net neutrality rules state the following *"enable the provision of specialized or innovative services on condition that they do not harm the open internet access. These services use the internet protocol and the same access network but require a significant improvement in quality or the possibility to guarantee some technical requirements to their end-users that cannot be ensured in the best effort open internet* [9]. *These specialized or innovative services have to be optimized for specific content, applications or services, and the optimization must be objectively necessary to meet service requirements for specific levels of quality that are not assured by the internet access services"*. The rules also urge that these services cannot be a substitute to Internet access service; and can only be provided if there is sufficient network capacity and therefore must not be to the detriment of the availability or general quality of Internet access service for end-users.

FCC's rules refer to above-mentioned service as non-Broadband Internet Access Service *(non-BIAS)* [1] *"Non-BIAS data services, which are not subject to the rules. According to the rules, non-BIAS data services are not used to reach large parts of the Internet, not a generic platform—but rather a specific "application level" service, and use some form of network management to isolate the capacity used by these services from that used by broadband Internet access services."*

8.2.4.2 The investment on the network infrastructure

The opponent of NN, which typically network operators argue that the NN regulation will make it more difficult for ISPs and other network operators to recoup their investments in broadband networks and so weaken the incentives to invest and upgrade the telecom infrastructure. Some ISPs have argued that they will have no incentive to make large investments to develop advanced fiber-optic networks if they are prohibited from charging higher preferred access fees to companies that wish to take advantage of the expanded capabilities of such networks [10]. FCC reclassified the BIAS as telecommunication service in the Open Internet Order [2]; FCC believed that the reclassification will preserve investment incentives.

8.2.4.3 The management of Internet traffic by Internet Service Providers and what constitutes reasonable traffic management

Commonly, "traffic management is used to effectively protect the security and integrity of networks. It helps to deal with temporary or exceptional congestion or to give effect to a legislative provision or court order. It is also essential for the certain time-sensitive service such as voice communications or video conferencing that may require prioritization of traffic for better quality. But there is a fragile balance between ensuring the openness of the Internet and the reasonable and responsible use of traffic management by ISPs" [11]. The opponents of NN argue that NN may prove ineffective in such a dynamic framework nowadays, leading to welfare-loss caused by congestion problems, arguing in favor of the possibility of differentiation of data packets according to their quality sensitivity [12].

EU urged that all traffic be treated equally but allow the network providers to make reasonable traffic management in consideration of justified technical requirements, so as to preserve the security and integrity of the network or to minimize temporary or exceptional network congestion. According to FCC rules, the no-blocking rule, the no-throttling rule, and the no-unreasonable interference/disadvantage standard will be subject to reasonable network management for both fixed and mobile providers of broadband Internet access service. Figure 8.3 shows a comparison between EU and the United States rules. We can see that regulators paid attention to the innovation when making their own NN policies. They leave space for the innovative services together with strict constraints.

8.3 CONASENSE

In the past 10 years, wireless and Internet technology have created an explosive growth in ICT services, supplying a wide range of personal and group data services. The limited data transmission capacity, as the bottleneck in the beginning, was broken little by little. The same wireless channel can now support much higher data rates and thus many more demanding services. There are also now a rapidly increasing demand and innovative application areas for

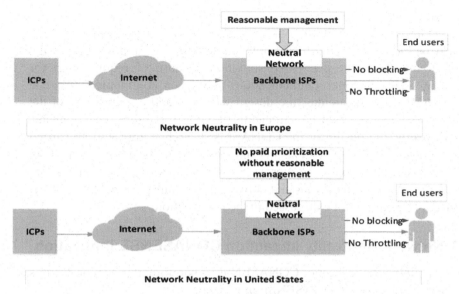

Figure 8.3 Comparison of NN between the USA and the EU [8].

services related to positioning, tracking and navigation. Meanwhile, sensing technology, sensors, and sensor networks have experienced an unprecedented development in the past several years. A variety of sensors types are now available in the market in many domains. And new sensor types are continuing to come out in different domains. As the integrated provision of these services will obviously raise people's living efficiency and QoL, and the ICT network can support much higher data transmission rate, it seems to be the right time to develop integrated CONASENSE services. One problem is that traditional approaches may not be optimal. Different services may have different frequency bands, waveforms, and hence different receiver platforms.

Most promising CONASENSE services may be available in 5–20 years. The new CONASENSE services should reflect the trend towards an information society in which applications and services become important likewise, bearing in mind that computing and communications should be integrated so as to save energy, software-defined radio combined with cognitive radio technology becomes increasingly important for new developments.

In order to achieve the goal of CONASENSE service, it is important for researchers and developers to identify the requirements for energy, terminal/platform, and receiver/system design concerning diverse application areas, such as e-health, security/emergency services, traffic management and control, environmental monitoring and protection, and smart power grid. They should pay much attention to the novel CONASENSE architecture design so as to minimize the energy consumption because of requirements for mobility, high data rate communications and green communications. The novel CONASENSE architecture design will help address problems of the existing architectures and be sufficiently flexible for future developments. Consequently, the design of the CONASENSE architecture will be carried out so as not only to integrate existing and novel communications, navigation, and sensing services but also to provide smooth transition between existing and new systems in hardware and software. NN impact on CONASENSE service innovation will be discussed in the next section.

8.4 Network Neutrality Impact on CONASENSE Innovation

QoS is an important aspect of CONASENSE. Therefore, this section focuses on QoS and discusses CONASENSE in general.

(A) From technology aspects, no discrimination requirement of Network Neutrality may have some negative impact on the new CONASENSE service innovation. The network nowadays is not neutral (Differentiated Network), it can supply QoS to different applications according to their characters and requirements. But, according to the new NN rules, QoS measures will be taken as discrimination, and be banned in the pure neutral network, as shown in Figure 8.4. It is obviously lowering the network efficiency and will cause congestion easily.

In the short term, NN will certainly make Internet content companies have more innovation space, encourage more innovative applications, and increase the efficiency of the society. But, in the long run, NN will inevitably weaken the enthusiasm of the investment on the network construction; it will lower the network quality level gradually. This in turn will influence Internet companies in the end. The reasons are as follows:

1) NN will impact the quality of the Internet service because NN limits the ability to guarantee the network QoS. As of now, there are mainly two types of QoS models [13].

- Integrated Services (IntServ) use Resource Reservation Protocol (RSVP) for signaling to invoke a pre-reservation network resource and traffic handling. IntServ can provide end-to-end guarantee for services and applications. But because Intserv is expensive and time-consuming, it has not been widely used in the Internet.

- Differentiated service (Diffserv) is a mechanism that identifies and classifies traffic in order to determine the appropriate traffic handling mechanism. It can integrate the same type of services and manage them together. It is now used widely.

To exemplify the models of QoS, we can imagine an accident site, there are several wounded persons: some of them got severely injured and need to go to hospital immediately, while others do not need so urgent actions. When the ambulances come, the doctors do not do much diagnosis and ask the severely wounded to get on the ambulance which makes that some non-significantly

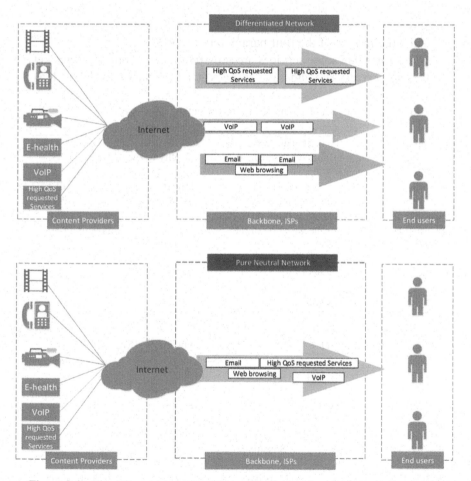

Figure 8.4 The difference between differentiated network and pure neutral network.

wounded cannot get on the ambulance in time. This case is a so-called Best-Effort Model (no QoS). The IntServ model is that, for all wounded whoever is serious or not, it would be necessary to make reservations for an ambulance; if the severely wounded did not make a reservation, he will not be sent to hospital. In the Diffserv model, when the ambulances come, the doctors will distinguish the level of wound and ask seriously wounded persons with similar situation to get on and be treated. So, we find Diffserv is an effective method for application and service transmission on internet. But with NN, differentiated service is banned in the public network.

2) Different Internet services and applications have different requests. QoS uses four parameters to judge the services request and include: bandwidth, delay, delay variation, and packet loss rate [14]. Table 8.1 presents the definition and the impact of the four parameters of QoS.

Table 8.2 shows that some services are sensitive to delay but not to packet loss, and some services are sensitive to packet loss but not to delay. It is

Table 8.1   Definition and impact of the four QoS parameters [14]

Parameter	Definition	Impact on QoS
Bandwidth	The maximum data that can be transmitted per second in the network.	It is to measure the transmission capacity of the network
Delay	The transmission time a service packet takes from one node to the other one	If the delay is too long, it will lower the QoS
Delay variation	It means the variation of the packet delays in the same flows	It is a key factor impacting QoS
Packet loss rate	It means the rate of loss of the data packets lost in transmission	Low packet loss rate will not impact the QoS

Table 8.2   QoS parameters for some applications [15]

Type	Bandwidth	Delay	Delay Variation	Packet Loss Rate
E-mail	Low	Not sensitive	Not sensitive	Not sensitive
WhatsAPP	Medium	Medium sensitive	Medium sensitive	Not sensitive
Video application	High	Sensitive	Sensitive	Sensitive
E-commerce	Medium	Sensitive	Sensitive	Sensitive
IoT, Industry 4.0	High	Sensitive	Sensitive	Sensitive
CONASENSE services	Large	sensitive	sensitive	Not known yet

necessary for a network to treat packets belonging to different applications differently in terms of their different requests. For instance, a network should give low-delay service to VoIP packets, but best-efforts service to e-mail packets [15]. Different services also need different QoS provisioning.

(B) NN may reduce the incentive of investment on the network construction. As we know, the funds usually flow to the market with big profits. If there is not enough profit from telecom network, investment on the network will certainly be reduced. Internet companies may also lose their interest to improve the efficiency of transmission, because they do not need to pay for the overuse of bandwidth. Insufficient investment on network with the overuse of bandwidth will certainly result in network congestion and inefficiency.

(C) The large amount of CONASENSE services, to some extent, can be taken as environment-sensitive services. And for people on air, on vessel, or on car, on road, their environmental situation including location, speed, temperature, health condition etc., are constantly changing. All the relevant information should be received correctly and timely for a CONASENSE service needed in order to make a correct and in-time decision, to help improve people's QoL. NN rules put relatively strict criteria of specialized services or no-BIAS services, sometimes on a case-by-case base. That may slow down the CONASENSE services experimenting, developing, and commercializing process.

High data transmission capacity is the base for the CONASENSE services. Most of the CONASENSE services focus on the future, on the assumption of Wireless Innovation System for Dynamically Operating Mega Communication (WISDOM) [16] or 6G network deployment, and extensive use of sensors. The value of CONASENSE services will not be realized on a congestive network without operational guarantees.

8.5 Conclusions

NN rules, its relevant debates, and its impact on ISPs and ICPs are discussed in this chapter as a background study. In this discussion, NN rules such as NB, NT, and NPP were covered including their exemplified definitions and impacts on present communication paradigm. These rules are user centric and may hamper the financial benefit of ISPs. The prohibition on any control in the flow of information through a communication network may refrain ISPs from earning quality-based revenues. In such case, ISPs may not be willing to enhance or upgrade the infrastructure to improve the QoS of the network. Therefore, although NN rules are beneficial for innovations currently, as campaigned by NN proponents, provided by NN rules, the end user may not "really" enjoy the QoS.

With the convergence of Internet and telecommunication network, basic telecommunication services have been moved to the public Internet network. Many of them have a quite strict QoS requirement so as to guarantee the relevant service responsibilities, such as emergency call and the basic service quality level. In this process, the policy maker and regulator should be careful to handle between service innovation, lawful customer rights and social responsibility. QoS models have enough reasons to remain as legal network functions.

We have investigated how NN rules, in their present form, will impact CONASENSE service in the future. Through some assumed futuristic scenarios, it is discussed that NN rules may not be favorable for innovation service. Banning Diffserv may put innovation service in the category of a usual communication system and ignore the very high data demand of the CONASENSE service. As customers may not welcome the innovative but poorly served new technology, the aspirant companies will struggle a capturing market for this innovative technology. Further, NPP may not allow users to choose better services among the choices offered by ISPs.

NN policy may stimulate the development of service innovation in short term but may be not good for the network base in long term, which may in turn hinder the service innovation, such as CONASENSE services. The relatively strict criteria for specialized services or non-BIAS services may slow down the CONASENSE's experimenting, developing, commercializing process.

Regulators are suggested to make careful decisions on NN policy to guarantee the long-term benefit according to their own situation. All the stakeholders should be encouraged to find on their own a way to ensure the prosperous development of the industry in the market.

References

[1] FCC, Category "For Consumers " "Open Internet", 2015.
[2] FCC, "The Open Internet Rules and Order" FCC 15–24, March, 2015.
[3] Ligthart, L., and Prasad, R. (2014). *Communication, Navigation, Sensing, and Services (CONASENSE)*. (Denmark: River Publisher).
[4] Janssen, C. (2015). "Over the Top Application (OTT)," *Techopedia* http://www.techopedia.com/definition/29145/over-the-top-application-ott

[5] Maxwell, W., Parsons, M., Farquhar, M. (2015). Net Neutrality–A Global Debate, Hogan Lovells Global Media and Communications Quarterly, 15–17.

[6] European Paralimentary "Our commitment to Net Neutrality", EU Actions, October, 2015.

[7] Wu, T. (2003). Network Neutrality, broadband discrimination. *J. Telecommun. High Technol. Law* 2, 141.

[8] FCC, "The Open Internet Order" FCC 10-201, December 21, 2010.

[9] European Commission, "Net Neutrality challenges", October, 27, 2015.

[10] Shin, D.-H., and Kim, T.-Y. "A web of stakeholders and debates in the network neutrality policy: a case study of network neutrality in Korea".

[11] European Commission, "Roaming charges and open Internet: questions and answers" 27 October 2015.

[12] van Schewick, B. (2015). Network neutrality and quality of service: what a nondiscrimination rule should look like? *Stanford Law Rev.* 67, 1, January.

[13] Fgee, El-B., Kenney, J.D., Phillips, W.J., Robertson, W., and Sivakumar, S. (). "Comparison of QoS performance between IPv6 QoS management model and IntServ and DiffServ QoS models" in: *The 3rd Annual Communication Networks and Services Research Conference*, 0-7695-2333-1/05, 2005, IEEE.

[14] de Gouveia, F. C., and Magedanz, T. Quality of Service in Telecommunication networks. *Telecommun. Systems Technol.* II.

[15] Hua Wei Technologies Co. Ltd. (2013). "QoS Technology White Paper", http://e.huawei.com/us/marketing-material/onLineView?MaterialID=%7B3623FE01-3572-4413-A71B-EBEBE9F2E141%7D

[16] Prasad, R. (2015). *5G Revolution Through WISDOM. Wireless Pers. Commun.* 81, 1351–1357, March.

Biographies

Y. Wang, she obtained Master's Degree from Beijing University of Post and Telecommunication in 2008. Currently, she is a guest researcher in CTIF in the field of Network Neutrality. Before 2008, she worked in Teleinfor Institute of CATR as a researcher and the field is telecommunication regulation, policy, and market. Till now, she works in International Cooperation Department of CATR, and is responsible for EU projects and ITU-D issues in China. The projects include:

- 2011–2012 Promotion of Green Economic Growth by Broadband Network,
- 2012–2013 Open China ICT Project-Observation of the Chinese telecom' development,
- 2012–2013 Implementing and planning outline of 'Smart Qianhai's policy (Qianhai is a region of Shenzhen),
- 2012–2014 International standard assessment of China Unicom from 2010 to 2010,
- 2014 WTDC: China Reception, Editorial Committee, Election for Mr. ZHAO Houlin,
- 2014 PP: China Reception, Editorial Committee, Election for Mr. ZHAO Houlin,
- 2014 Application for ITU Centers of Excellence (CoE) in Conformance and Interoperability (C&I),
- 2014 Application for Conformance and Interoperability Test lab.

Prof. R. Prasad has been holding the Professorial Chair of Wireless Information and Multimedia Communications at Aalborg University, Denmark (AAU), since June 1999. Since 2004, he is the Founding Director of the Center for TeleInfrastruktur (CTIF-http://www.ctifgroup.dk/), established as a large cross-/multi-disciplinary research center at the premises of Aalborg University. Under the leadership of Ramjee Prasad, CTIF has emerged as a prominent international center of excellence for his visionary ideas and path- breaking research in wireless communications. CTIF boasts of a global presence today with its divisions dotted in eleven coutries (and counting) spanning across three continents. As well as, through the numerous valuable partnerships, it has forged with world renowned academic intitutions spread across fourteen countries and six continents.

He is a Fellow of the Institute of Electrical and Electronic Engineers (IEEE), USA, the IET, UK, the IETE, India, the Wireless World Research Forum (WWRF), and a member of the Netherlands Electronics and Radio Society (NERG), and the Danish Engineering Society (IDA).

For his exceptional contribution to the internationalization of the Danish telecommunication research and education, in 2010, Ramjee Prasad was awareded the Knight of the Order of Dannebrog (Ridderkorsetaf Dannebrogordenen-2010) by the Queen of Denmark.

Ramjee Prasad is the recipient of many international academic, industrial, and governmental awards and distinctions. He has received many prestigious international awards such as: IEEE Communications Society Wireless Communications Technical Committee Recognition Award in 2003 for making contribution in the field of "Personal, Wireless and Mobile Systems and Networks", Telenor's Research Award in 2005 for impressive merits, both academic and organizational within the field of wireless and personal communication, 2014 IEEE AESS Outstanding Organizational Leadership Award for "Organizational Leadership in developing and globalizing the CTIF (Center for TeleInFrastruktur) Research Network", and so on.

He received a honorable award with Gold Medal in 2014 from the Academic Council of Technical University-Sofia, Bulgaria, for his major contribution to the development of its international cooperation.

He has published more than 30 books, 1000 plus journals, and conferences publications, more than 15 patents, over 100 Ph.D. Graduates, and larger number of Masters (over 250). Several of his students are today worldwide telecommunication leaders themselves. His research publications have been very well cited globally.

9

CONASENSE at Nanoscale: Possibilities and Challenges

Prateek Mathur[1,2], Rasmus H. Nielsen[1], Neeli R. Prasad[1]
and Ramjee Prasad[1]

[1]Center for TeleInFrastruktur, Aalborg University, Aalborg 9220, Denmark
[2]Global ICT Standardization Forum for India, New Delhi, India

9.1 Introduction

Communication, Navigation, Sensing, and Services (CONASENSE) is an approach targeting to determine and examine the possible services and applications that can benefit by considering them collectively. Emergence of nanotechnology is changing the existing landscape across several disciplines in the academia and business proposition in the industry across several verticals. Nanotechnology has opened up a new avenue with the limitless possibilities to look at the nanoscale. The major thrust areas within nanotechnology are detailed out in this chapter stating the possible applications of CONASENSE, the state-of-the-art, and the possible challenges that have to be addressed to achieve the objectives. Nanotechnology was introduced by Feynman [1] in his lecture in 1995 "There's Plenty of Room at the Bottom". Broadly, the nanotechnology has been looked at in terms of nanoparticles, nanofluids, nanomaterials, nanomachines, and nanorobots (nanobots). Nanoparticles can be referred as basic building blocks of nanoscale world. Nanomachines and nanorobots refer to nanoscale devices that are expected to replicate and function as dedicated operational units with the predefined purpose. Nanofluids refer to a mixture of nanoparticles along with a base fluid to function for a given application. Broadly, there are two approaches in nanotechnology that are adopted, i.e., top–down and bottom–up approaches (nanotechnology in concrete), shown in Figure 9.1. In the top–down approach it is considered that the larger structure are reduced in size

Role of ICT for Multi-Disciplinary Applications in 2030, 185–200.

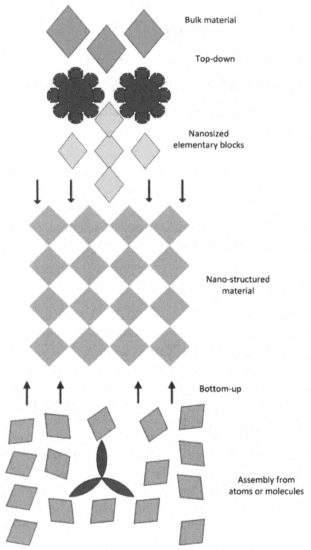

Figure 9.1   The top–down and bottom–up approaches (adapted from Sanchez and Sobolev [2]).

(miniaturization), whereas in the bottom–up it is where materials are made from atoms/molecules. The bottom–up approach is referred as molecular manufacturing, till date mainly the focus has been on top–down approach. This chapter details out the current status in terms of communication, sensing, and navigation at the nanoscale. Subsequently, the possible applications based on them are detailed out, and the detailing is done also for the key challenges that are to be addressed for their implementation. Since a comparison is made between various operations at the nanoscale and the operations visible to naked eye, throughout this paper they are referred with the word normal scale.

9.2 Communication at the Nanoscale

The communication at the nanoscale has been broadly proposed mainly in two types that is electromagnetic communications and molecular communications. The electromagnetic communication systems rely on very high frequency (VHF) and ultra-high frequency (UHF), but the prominent focus is on using the Terahertz frequency band for the nanoscale applications. Utilizing electromagnetic communication relying on terahertz frequencies the concept of nanonetworks and nano-Internet of Things ((IoT) has been stated in the literature. Wherein some nano-devices function as the nano-routers, while other function as the sensing devices. The utility of such a nano-IoT or nanonetwork looks promising especially in regard to a body area network at the nanoscale. The operation of IoT at the nanoscale – nano-IoT has been presented in Akyildiz and Jornet [3] relying on the terahertz band. Two possible application areas are proposed to describe the working operation of nano-IoT, i.e., intrabody network and interconnected office. Similar to the normal-scale IoT, nanonodes are considered as the things that would do the sensing/monitoring at the nanoscale. The nanonodes communicate through nano-routers, and nano-micro interfaces that club the information from diverse routers transmitted back and forth to microscale devices. Finally, there are the gateway devices that relay this information over the Internet relying suitable communication mechanism as used for the normal-scale IoT. This process has been shown in Figure 9.2.

The communication at the nanoscale necessitates formulating the functioning of the various layers of the communication system, since the protocols for the normal-scale operations are too complex for direct application. The modulation techniques, channel modeling and path loss and propagation patterns are some of the key factors to be looked into at the physical layer. Authors Jornet and Akyildiz [4] have demonstrated a transceiver for communication in this band utilizing a high-electron mobility transistor built with a boron-based (III–V) semiconductor enhanced by using graphene. The possible channel sharing mechanisms, format of information exchange (packet structure and size), addressing scheme for the nodes

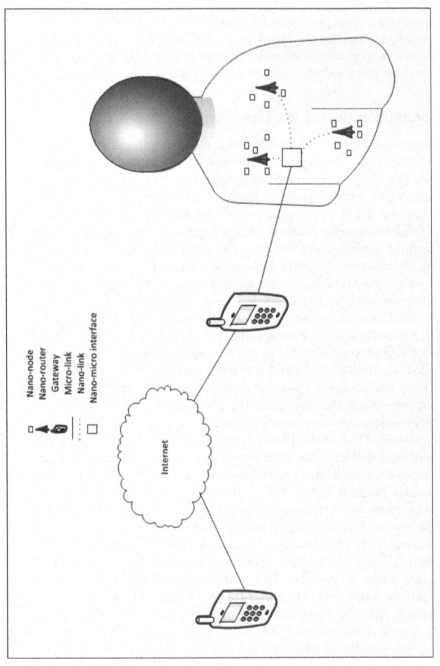

Figure 9.2   Nano IoT operation for intra-body network (adapted from Akyildiz and Jornet [3]).

are some of the issues relating to the other upper layers. Similarly, a key issue would be: "how would the communication be secured and at which layer is this functionality going to be provided?"

Another major hurdle is the availability of energy resources, as the availability of direct energy supply using a nano-battery onboard the nano-device would not be capable to support resource intensive operations. An alternative could be relying on energy harvesting mechanisms, but they will also have certain operational and generational capacity constraints. However, between onboard battery and harvesting solutions, it looks the energy harvesting solutions have an upper hand. The authors in Hu et al. [5] present a nanogenerator that generates sufficient power to run a radio and communicate with the radio receiver placed 5 m away from the transmitter that is connected with the generator. Straining a cantilever beam is utilized as the source, with an applied strain of 0.12%, it was proven that an output voltage of 10 V and output current of 0.6 μA could be obtained. The volume power density is noted to be 10 mW/cm^3. Clearly, these would be promising for powering the nanoscale devices. A major limitation and point of concern for operation at nanoscale taking into account that new details for layer-wise operation have to be developed is the reliability of the nanoscale operations. This factor is even compounded by considering the environmental heterogeneity that prevails at nanoscale.

The molecular communication that is considered as an alternative approach to the electromagnetic communication relies on molecule to molecule interaction for conveying messages. This communication scheme takes a cue from the biological systems wherein hormones and small proteins, peptides, lipids are capable of carrying a message between a transmitter and receiver [6]. This is a highly specialized system, as the molecular receptor has to be capable of differentiating between the transmitters. Possible nanoscale entities that could function on similar mechanism could be used to detect and/or treat biological disorder (tumors) and diseases such as cardiovascular diseases. Therefore, it can be concluded that communication systems at the nanoscale are still immature and require further investigation and research before implementations at large scale could be feasible.

9.3 Sensing at the Nanoscale

Achieving sensing utilization similar to the manner in which sensor nodes are utilized in wireless sensor networks is not feasible at the nanoscale. This is due to the above stated major reason of a lack of active communication between the nanoscale devices. The idea of nanoscale motes with capabilities and functionalities similar to an usual sensor mote has been discussed in the literature, but it is the miniscule form size that is a major limiting factor to develop a mote at nanoscale. The authors in Chou et al. [7] show the utility of a wireless sensing-based information transmission concerning a glucose biosensor that has been modified using graphene and magnetic beads. The wireless sensor utility is shown in terms of a Zigbee module that transmits the input from a microcontroller connected to the biosensor solution to a Zigbee coordinator connected to a computer. This sensing unit does not have anything to do directly with the biosensor and nanoscale operations.

The majority of the nanoscale applications utilizing nanoparticles could be referred as passive sensing. The utility of nanoparticles for passive sensing has been enormous and is growing larger and larger. Utilization nanoparticles as Quantum Dots has been stated in Toumey [8] as a mechanism to detect the cancerous sites in the body with radiation illumination. Similarly, using a nanoparticle based mixture to study the contents of the soil based on interaction between the nanoparticle and soil nutrients has been reported in Baruah and Dutta [9]. Nanosensors can be used to determine the amount of microbes, and pollutants in the soil system concerning agricultural cultivation. Utilizing nanoparticles as doping agents to determine the presence of organic pollutants by carrying out a process of photo catalysis, wherein semiconducting oxides can break complex organic molecules by absorbing photons is an important area within the agriculture field that finds application of nanoparticles and nanotechnology [9]. Similarly, determining the possible existence of *Salmonella* in food products to prevent the spread of disease salmonellosis wherein the nanoparticles in form of quantum dots were used and the fluorescence response was measured with a fluorometer. The fluorescence response of the quantum dots changes with the change in the mixing with samples containing *Salmonella* [10]. Similar is the concept of enhancing the medical imaging using doping with nanoparticles and increasing the utility and performance of sunscreen creams [11].

Another possible utility type of nanotechnology has been in the form of nanorobots (referred commonly as nanobots), wherein the objective is that the nanobot relocates itself to a target location and accomplishes a certain specific task. A nanobot which is able to transfer miniscule molecular loads has been described in Douglas et al. [12]; it has a shape of a hexagonal barrel that has a lid that is opened/closed based on cell signaling. The closing and opening of the lid is based on aptamer, i.e., a molecule that recognizes cancer cells based on their surface. If the cancer cell is detected, the chemical lock-holding gate container dissolves and opens the container, thereby establishing the contact with contained molecules and the cancer cell, and then triggering its death. This is again an example of passive sensing by the nanoparticles to respond a particular manner based on the environmental information. This nanoparticle-based barrel is also referred as a scaffolding structure, and in future nano scaffolding structures could be used for engineered tissues, that could in turn be used for detecting and sensing variety of medical issues. Similarly, use of nanoscale polymer functions as membranes, fibers within the body which contract or expand by applying external heating or cooling. Indirectly, the application of external heating/cooling and a response by the material is its capacity to detect the external influence and change accordingly [11].

The passive sensing and actuation of the nanoscale devices and utilities are clearly illustrated in the above examples. Compared with an active sensing mote that could sense for physical parameters such as vibration, pressure, and temperature and report it to distant nodes relying on communication in a certain predefined manner active sensing is still considered as a distant reality.

9.4 Navigation at the Nanoscale

Compared to communication and sensing discussed in the previous sections, navigation is even more challenging at the nanoscale. Navigation requires an understanding of the environment that includes details of possible landmarks or localization beacon/anchor points that would help to follow a desired path. In classical navigation issues the indoor localization is an unresolved problem in comparison to the outdoor localization that relies usually relying on the global navigation satellite system, e.g., the global positioning system (GPS). The major problem with indoor localization is the environment limits the reception of the GNSS signal in indoor environments. Alternatively, in the indoor environment localization and possible navigation are experimented relying on received signal strength, time-of-arrival (ToA), angle-of-arrival (AoA), time-difference-of-arrival (TDoA) etc., of the radio signals (WiFi, Bluetooth, LTE, etc.). Subsequently, by applying principles of triangulation the location can be determined [13]. Alternatively, utilizing a fingerprint approach wherein the signal samples are stored in a database and along with the coordinates of the signal, or using a crowd-sensing approach wherein people volunteer to share their information gathered by smartphone sensors we are able to build indoor environment maps [14]. The purposes of localization and navigation require different amount of accuracies for localization; coarse localization accuracy is acceptable for human navigation but not for other specific applications [14].

In the case of nanoscale navigation and localization the situation becomes even more complex as in comparison to indoor localization. The internal structure and layout will vary drastically from nanomaterial/nanofluid and the environment in question. Consider for example the case of internal structure of human body and internal combustion engine. Additionally, as already stated the communication between the entities at nanoscale amongst themselves and with entities at normal-scale in the adjoining environment required to be addressed first, as also discussed earlier. Based on this it can be inferred the amount of complexity involved in navigation and guidance at the nanoscale. If the communication can be established between nanoscale devices and particles relying on IoT (device for talking in between) and there is the possibility for sensing to take place through the nanoscale devices and by the associated normal-scale material/devices and entities, it should become possible for navigation to take place.

Utilizing an external force applied through a permanent magnet to move a metallic medicine through a mouse to a selected part of the body has been considered as a possible navigation mechanism. Similarly, use of ultrasound to direct sound waves onto bubbles carrying a payload (similar to the nanobot hexagon barrel stated earlier), shows that the bubbles can burst with sufficient force to push the payload into the tissue [15]. Both the approaches have serious limitations in terms of navigation with respect to accuracy to a specific location. Moreover, they require an external agent to influence the movement. The better way would be wherein the nanoscale entities interact with their environment that could comprise entities at the usual normal-scale and then collectively decide on a relocation/navigation strategy, with the human involvement probably limited just to approve the decision or to fine-tune it. Utilizing an external medium to move the fluid similar to the case of fluid movement relying on peristaltic pumping can move fluids through smooth muscle tubes in the body, e.g., bile duct, gastrointestinal tract, and esophagus [16]. The nanofluids have been utilized to evaluate the influence under the establishment of conduction and convection cycles and as a heart transfer liquid. But the utility of navigating the nanofluid specifically through a predefined path has not been explored.

9.5 Possible Applications of CONASENSE at Nanoscale

The possibilities of enabling effective communication, sensing, and navigation can have a major impact and can bring a fundamental changes in the overall functional world of today. Similar to the transformation that sensing and sensor networks can do at the normal-scale, the nanoscale devices can reach and have far more pronounced influence. The envisioned idea of having a smart grid connecting the power generation and supply in an intelligent manner with the household demand and industrial demand could be transformed with nanotechnology. Smart metering wherein the users could adjust their usability for the household chores based on variable electric supply cost, e.g., laundry could be done when the offer rate for power supply is lower [17]. Some of the possible envisioned applications are discussed in the following text. In addition to this, when utilizing nanotechnology within the washing machine, the devices could talk among themselves and alert the home user of a water pump that is not working at full efficiency, or that the rotating efficiency has declined due to a worn belt for the motor. Similarly, the cooling fluid in the power supply equipment such as transformers could have nanofluid to help determine about the internal functioning of the equipment with nanoparticles also embedded in the components that form the transformer. Considering that with aging and continuous internal operation, the properties of the components would change in terms of texture, strength, etc. Similarly, nanofluids could be injected directly into the stem of plant/trees or along with irrigation water through the roots, and by storing the evaporated water from the leaves, it could be determined as to how well is the internal system/physiology of the plant. This can in turn be utilized for determining the appropriate action required to fix appropriate nutrients (fertilizers and manure) and utilization of insecticides and pesticides in the appropriate quantity and form. Improper use of these mechanisms has been observed to have drastic influence on the plant physiology and in result the crop output. The nanobots could move through the possible vacant spaces in constructed structures such as bridges, rails, dams, and walls. The navigating nanobot could communicate much more precise details of where is the possible fault with the external information/medium gateway. This could even be applied to engineered structures such as aircrafts, and transportation vehicles. The possible applications of nanoscale in concrete have been looked in great detail [2]. Similarly, the possibility of clothes embedded with nanoparticles to determine whether there is a need for laundry or it could be delayed for a while. Utilizing the nanoscale entities it could be possible to determine the quality and reliability of the raw material that go into making a given commodity. Similar to the manner stated earlier regarding the home appliances and other machines the various machines and equipment that interact with each other could have embedded nano-entities that could determine and report the details to external devices and environment.

Since there would be a plethora of nano-devices, it would be even possible to identify an individual automatically, say a person walks into a shop that sells clothing merchandise and has a membership system. In the usual case the person would present a membership card number that identifies him/her; with nanotechnology the individual could be identified through the worn clothes. This would be possible with nanoparticles in the clothes talking with the communication equipment at the store. Based on this example, it is evident, that such an implementation could revolutionize the complete business models for retail customer industry. In a way, such information exchange would lead to the operation that were routed as the future web, referred as "semantic web", that transforms the manner in which the communication takes place on the internet to serve the world wide web, relying on the use of semantic connotations and web syntax that supports it.

The possible use of nanotechnology in medical sciences is enormous; indeed the most important area wherein the possibilities of nanotechnology are being investigated is medical sciences. This gains special prominence since most of the complex unresolved issues of medical science in terms of medical disorders and diseases require involvement at the microscale and the nanoscale, in terms of examining the functioning of specific cells and molecular compounds in the biological system. The authors Mesiti and Balasingham [18] present a nanomachine that would be capable of direct interactions with the neurons on the input side referring it as synaptic nanomachine. Utilizing nanomachines for delivering drug to a specific cell in the human body is being actively researched as a possible way to attack the cancerous tumor. The possibilities of carrying out a nanosurgery utilizing multiple nanomachines that coordinate collectively and that carry out the tasks by operating instruments and actuations required has been discussed in Song et al. [19]. The nanomachines are, however, controlled externally by a human being and therefore they are not completely autonomous. Similarly, the use of an atomic force microscopy (AFM)-based nanorobot for analysis, imaging and tracking capabilities for carryout out surgeries at the cellular level has been presented in Freitas [20]. Similar to the previous application of nanosurgery, the nanorobot is controlled through AFM and therefore the nano entity is externally controlled and requires human intervention. Nanosensor and entities could be injected into animals similar to the manner in which they could be used for medical utilities for human beings. In addition to monitoring the wellbeing of animals the nano-entities would help in detecting the movement of animals in a more reliable manner and also make it almost infeasible for the poachers and smugglers to kill an animal and get away without getting noticed.

Based on the above possible applications, it can very well be said that nano-technology implementation wherein there are its capacity of communication, sensing, and navigation would provide useful data that would be manifold higher compared to that envisioned with the world moving towards massively 5G-based communication, cloud-based storage, and IoT-based interaction in the ecosystem. This is, however, underpinned by the assumption that all the data communication between the nanoscale entities and the interface with the normal-scale devices is transmittable over the internet. This is a fair assumption considering the fact that all possible communication technologies on the normal-scale allow data transmission onto the internet using suitable interfaces and data conversions.

9.6 Major Challenges

The utility of nano entities is enormous as demonstrated above in terms of possible communication, sensing and navigation. However, there are some serious challenges that will have to be addressed before allowing for a large scale implementation of the nano entities based utilities. Implementation of nanoscale entities could invade the privacy of individuals. Similarly, secrecy invasions can take place for nano-based implementation in operations of organizations, design secrets for equipment/machines, which will be highly undesirable for the business economy to flourish.

Similarly, implementation of nanotechnology based utilities could have an ill effect on health. Especially, utilizing them for medical purposes, and even use in animals and agriculture require to be carefully looked into, since both could influence the wellbeing of the living life in form of animals and plants and in-turn also influence the lives of human beings. Uncontrolled use of nanotechnology could have a detrimental effect on the internal functioning of body system, and could lead to emergence of new medical ailments or diseases. Another major concern with implementation of nanotechnology is the nanoscale that prohibits any activity to be captured directly by the eye, special equipment is necessitated to see the implementation and operation of the nanoscale device. This could be a major hindrance in large-scale implementation, as the possible monitoring and checking would be complex. Additionally, large-scale implementation of nanotechnology could prove to be a major boom in the artificial intelligence implementation, and raise the risk associated with artificial intelligence, i.e., machines dominating human beings. There is also an accounted risk of environmental pollution and degradation with utilization of nanotechnology. These concerns necessitate the need for clearly defined public policies that detail out the various possibilities in the nanotechnology ecosystem that are permissible and the procedure to use the technology, and any applicable mitigation measures.

9.7 Conclusion

The possible utilities of bringing CONASENSE to the nanoscale are enormous, subject to overcoming the major challenges that are yet to be addressed. Establishment of a possible communication and interaction system that can operate across different operational environments would be a major milestone in achieving an effective utility in terms of services and applications of communication, sensing, and navigation at the nanoscale. The possible utility with the top-bottom approach is the one stressed and discussed in this chapter. The utility of bottom–up approach could be even more promising and could open possible avenues that cannot be even imagined at the moment. It would be appropriate to say that "There's Plenty of Room at the Bottom" at the nanoscale, as stated by Richard Feynman, was absolutely well justified, even in terms of possibilities for CONASENSE.

References

[1] Feynman, R. P. (2016). *There's Plenty of Room at the Bottom*. Available at: http://www.zyvex.com/nanotech/feynman.html [accessed 22 Feb 2016].
[2] Sanchez, F., and Sobolev, K. (2010). Nanotechnology in concrete: a review, construction and building materials. *Constr. Build. Mater.* 24, 2060–2071. doi: 10.1016/j.conbuildmat.2010.03.014
[3] Akyildiz, I. F., and Jornet, J. M. (2010). The Internet of nano-things. *IEEE Wirel. Commun.* 17, 58–63.
[4] Jornet J. M., and Akyildiz, I. F. (2014). "Graphene-based plasmonic nano-transceiver for terahertz band communication," in *2014 8th European Conference on Antennas and Propagation (EuCAP)*, The Hague, 492–496. doi: 10.1109/EuCAP.2014.6901799
[5] Hu, Y., Zhang, Y., Xu, C., Lin, L., Snyder, R. L., and Lin Wang, Z. (2011). Self-Powered System with Wireless Data Transmission. Nano Lett. 11, 2572–2577.
[6] Felicetti, L., Femminella, M., Reali, G., and Liò, P. (2016). Applications of molecular communicatio ns to medicine: a survey. *Nano Commun. Networks* 7, 27–45. doi: 10.1016/j.nancom.2015.08.004

[7] Chou, J. C., Chen, R. T., Liao, Y. H., Chen, J. S., Huang, M. S., and Chou, H. T. (2015). Dynamic and wireless sensing measurements of potentiometric glucose biosensor based on graphene and magnetic beads. *IEEE Sensors J.* 15, 5718–5725.

[8] Toumey, C. (2013). Nanobots today. Nat. Nanotechnol. 8. doi: 10.1038/nnano.2013.128

[9] Baruah, S., and Dutta, J. (2009). Nanotechnology applications in pollution sensing and degradation in agriculture: a review. *Environ. Chem. Lett.* 7, 191–204.

[10] Kim, G., Moon, J.-H., Moh, C.-Y., and Lim, J.-G. (2015). A microfluidic nano-biosensor for the detection of pathogenic *Salmonella. Biosensors Bioelectron.* 67, 243–247.

[11] Roco, M. C. (2013). Rise of the nano machines. *Sci. Am.* 308, 48–49.

[12] Douglas, S M., Bachelet, I., and Church, G. M. (2012). A logic-gated nanorobot for targeted transport of molecular payloads. *Science* 335, 831–834.

[13] Dardari, D., Closas, P., and Djurić, P. M. (2015). Indoor tracking: theory, methods, and technologies. *IEEE Transact. Vehic. Technol.* 64, 1263–1278. doi: 10.1109/TVT.2015.2403868

[14] Zhang, C., Subbu, K. P., Luo, J., and Wu, J. (2015). GROPING: Geomagnetism and cROwd sensing Powered Indoor NaviGation. *IEEE Transact. Mobile Comput.* 14, 387–400. doi: 10.1109/TMC.2014.2319824

[15] Greenemeier, L. (2015). Launch the nanobots. *Sci. Am.*, 312, 50–51.

[16] Tripathi, D., and Bég, O. A. (2014) A study on peristaltic flow of nanofluids: application in drug delivery systems. *Int. J. Heat Mass Transf.*, 70, 61–70. doi: 10.1016/j.ijheatmasstransfer.2013.10.044

[17] Kayastha, N., Niyato, D., Hossain, E., and Han, Z. (2014). Smart grid sensor data collection, communication, and networking: a tutorial. *Wirel. Commun. Mob. Comput.* 14, 1055–1087. doi: 10.1002/wcm.2258

[18] Mesiti, F., and Balasingham, I. (2013). Nanomachine-to-neuron communication interfaces for neuronal stimulation at nanoscale. *IEEE J. Select. Areas in Commun.* 31, 695–704. doi: 10.1109/JSAC.2013.SUP2.1213002

[19] Song, B., Yang, R,, Xi, N., Patterson, K. C., Qu, C., Lai, K. W. (2012). Cellular-level surgery using nano robots. *J. Lab. Autom.* 17, 425–434. doi: 10.1177/2211068212460665

[20] Freitas, R. A. (2005). Nanotechnology, nanomedicine and nanosurgery. *Int. J. Surg.* 3, 243–246. doi: 10.1016/j.ijsu.2005.10.007

Biographies

P. Mathur is currently enrolled as a PhD student at Center for TeleInFrastruktur (CTIF) at Aalborg University (AAU), Denmark. He received his MSc degree from University of Bradford, UK and Bachelor Of Engineering from University of Pune, India in the year 2009 and 2008, respectively. His research focus is on effective utilization of mobile nodes in wireless sensor networks.

R. H. Nielsen is an assistant professor at Center for TeleInFrastruktur (CTIF) at Aalborg University (AAU), Denmark and is currently working as a senior researcher at CTIF-USA, Princeton, USA. He received his MSc and PhD in electrical engineering from Aalborg University in the year 2005 and 2009, respectively. He has been working on a number of EU- and industrial funded projects primarily within the field of next generation networks where his focus is currently security and performance optimization. He has a strong background in operational research and optimization in general and has applied this as a consultant within planning of large-scale networks. His research interests include IoT, WSNs, virtualization, and other topics related to next generation converged wired and wireless networks.

N. R. Prasad is an IEEE Senior Member, Head of research at the Center for TeleInfrastruktur (CTIF) at Aalborg University and Director of CTIF-USA, Princeton, USA. She has over 14 years of management and research experience both in industry and academia. She has gained a large and strong experience into the project coordination of EU-funded and Industrial research projects. Her current research interests are in the area of QoL, SON, IoT, Identity Management, mobility, network management, and monitoring; practical radio resource management; cognitive learning capabilities and modeling; Security, Privacy, and Trust. Experience in other fields includes physical layer techniques, policy based management, short range communications. Her publications range from top journals, international conferences and chapters in books. She has also co-edited and co-authored two books and has over 50 peer reviewed papers in international journals and conferences. She is also very active in several conferences as chair and as program committee member.

R. Prasad is currently the Director of the Center for TeleInFrastruktur (CTIF) at Aalborg University, Denmark and Professor, Wireless Information Multimedia Communication Chair. Ramjee Prasad is the Founding Chairman of the Global ICT Standardisation Forum for India (GISFI: www.gisfi.org) established in 2009. GISFI has the purpose of increasing of the collaboration between European, Indian, Japanese, North-American, and other worldwide standardization activities in the area of Information and Communication Technology (ICT) and related application areas. He was the Founding Chairman of the HERMES Partnership, a network of leading independent European research centers established in 1997, of which he is now the Honorary Chair. He is a Fellow of the Institute of Electrical and Electronic Engineers (IEEE), USA, the Institution of Electronics and Telecommunications Engineers (IETE), India, the Institution of Engineering and Technology (IET), UK, and a member of the Netherlands Electronics and Radio Society (NERG), and the Danish Engineering Society (IDA). He is also a Knight ("Ridder") of the Order of Dannebrog (2010), a distinguished award by the Queen of Denmark. He is the founding editor-in-chief of the Springer International Journal on Wireless Personal Communications. He is a member of the editorial board of other renowned international journals including those of River Publishers. Ramjee Prasad is a member of the Steering committees of many renowned annual international conferences, e.g., Wireless Personal Multimedia Communications Symposium (WPMC); Wireless VITAE and Global Wireless Summit (GWS). He has published more than 30 books, 900 plus journals and conferences publications, more than 15 patents, a sizeable amount of graduated PhD students (over 100) and an even larger number of graduated MSc students (over 200). Several of his students are today worldwide telecommunication leaders themselves.

Index

9 788793 379480